P9-AQH-880

WITHDRAWN

STORIES FROM SCIENCE

Stories from Science

BY A. SUTCLIFFE & A. P. D. SUTCLIFFE

Book 1

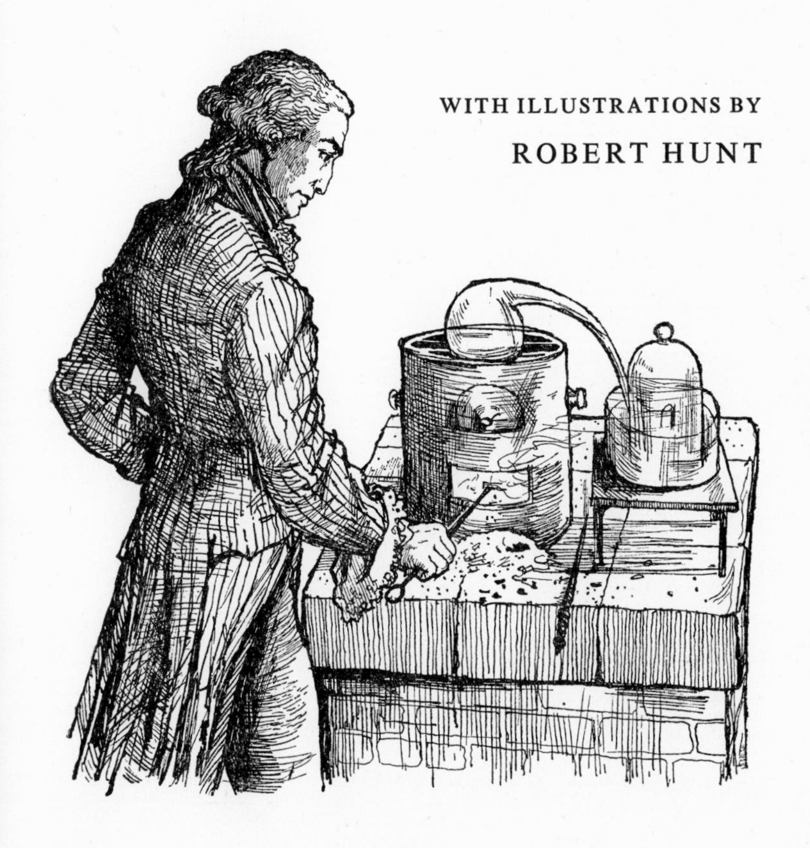

WITH ILLUSTRATIONS BY
ROBERT HUNT

CAMBRIDGE: AT THE UNIVERSITY PRESS

PUBLISHED BY
THE SYNDICS OF THE CAMBRIDGE UNIVERSITY PRESS
Bentley House, 200 Euston Road, London, N.W.1
American Branch: 32 East 57th Street, New York 22, N.Y.
West African Office: P.O. Box 33, Ibadan, Nigeria

Printed in Great Britain at
The Stellar Press, Barnet, Herts.

CONTENTS

PREFACE

One of the authors, when a young science master in Cambridge, decided to collect stories of unusual incidents or chance occurrences in science and engineering which would enrich his teaching and interest his pupils. So began a hobby of collecting stories which has given him much pleasure during the last forty years. With the help of his son, these stories have been prepared for publication in the hope that they will give similar pleasure to others.

It will be obvious that in order to gather all this information together, reference has been made to a variety of sources. A list of some of the books and articles consulted is given at the end of each volume, and the authors desire to express their sincere thanks to all those writers whose works have proved so helpful.

The illustrations, which add so much interest to the book, are the work of Mr Robert Hunt, who has successfully combined great accuracy of detail with artistic skill.

The authors gladly acknowledge their gratitude to many colleagues and friends: to Mr G. H. Franklin who translated numerous passages, and to Mr L. R. Middleton, Mr J. Harrod, Dr A. H. Briggs, Dr R. D. Haigh and Miss M. Lipman, who read the typescript.

They are particularly indebted to Mr R. A. Jahn for that severe yet constructive criticism which only a friendly colleague of many years' standing can give. In its final stages before printing the book owes a great deal to the officers of the University Press for their very helpful suggestions and emendations.

A. S.
A. P. D. S.

LINCOLN 1962

1. The Art of Making Glass

GLASS HAS been used by man for three or four thousand years at least, although it does not occur in nature. Long before the birth of Christ it was manufactured in the biblical country of Canaan, with the centre of the industry at the town of Sidon.

Canaan, or Phoenicia, as it was called by Pliny the Roman historian, was a flourishing district on the Mediterranean coast of Syria watered by the River Belus. This river rises in the marshy district near Mount Carmel and flows sluggishly on its short journey of five miles to the sea, carrying down much sand and soil which it deposits on a narrow half-mile strip of land at its mouth. The tidal waters of the ocean wash the sand daily, dissolving and carrying away many of the impurities, and leaving a silvery sand which sparkles in the sun.

This narrow sandy strip was the scene of the accidental discovery by the Phoenicians of the method of making glass.

The Phoenicians were an industrious race of people. Some of their menfolk were merchants who sailed the seas of the then known world and bartered their manufactured wares for raw materials. They visited Ancient Britain, for example, and traded their cloth for tin, mined in Cornwall. They frequently visited nearby Egypt, returning with a cargo of crude soda – that is, sodium carbonate (washing soda) mixed with small amounts of sodium bicarbonate, common salt and other impurities. Crude soda occurred plentifully on the sands by the soda-lakes of Egypt, and was used for washing clothes and other things; for in those days soap was unknown. The Egyptians also used it for embalming when preserving dead bodies as mummies.

The story of the accidental discovery is told by Pliny, who relates that sailors from a ship laden with lumps of crude soda landed on the narrow coastal strip near the Belus and began to prepare a meal. On this very sandy shore they could find no stones suitable for propping their cooking pot over the fire, so they used instead some of the lumps of soda from their ship. They lit their fire of wood beneath the cooking pot and in time, to their great surprise, they saw 'transparent streams flowing forth of a liquid hitherto unknown which had been formed by the action of the heat of the fire on the soda and sand. This transparent stream was liquid glass'.[1]

A glassy stream flows from the fire

The method of making glass by heating together soda and sand which had been discovered in this accidental way was then developed by the inventive and skilful Phoenicians, and soon they were producing objects of glass. The first of these were probably coloured beads which their merchant seamen could barter for useful materials with the less civilised peoples of other countries.

Much has been written about the truth or otherwise of this account of the discovery of glass-making. The story correctly states the main requirements for making glass – sodium carbonate, sand and heat. The sand of that particular strip of land near the Belus is of a most suitable kind for this purpose and, indeed, was used in glass manufacture for centuries. But many authors have doubted whether the heat from a wood fire on an open beach would reach a high enough temperature to fuse soda and sand together into liquid glass. The necessary temperature for this to take place depends largely on the composition of the mixture of sand and soda. A recent experiment has shown that a hot fire of wood burning for two hours in the open developed a temperature high enough to melt the mixtures used in making most kinds of glass.[2] There is, of course, no evidence that the fire of the Phoenician sailors gave out as much heat as the experimental one. Nevertheless, it would not be unreasonable to assume that the heat generated by it would be sufficient at least to give the surface of the mixture a glassy shine. This could have given an intelligent man a clue that a new, shiny, attractive substance could be produced by the deliberate heating together of soda and sand in the very hot fire of a specially made furnace.

The following different account of the discovery of glass mentions a very hot fire.

> Some say that the Children of Israel set fire to a forest and the fire became so fierce that it heated the nitre with the sand so as to make them melt and run down the slope of the hills and that thenceforward they sought to produce artificially what had been effected by accident in making glass.[3]

By nitre the author of this statement meant an alkaline substance, either soda or one chemically similar to it.

The possibility that the Phoenicians or the Israelites, or even both of them, discovered how to make glass for themselves, in the accidental manner told above, cannot be ruled out altogether, although there is very little evidence indeed to support the story about the Israelites. Nevertheless, it is well established that the method was known to the Ancient Egyptians long before either of these two races knew about it, for glass objects have been found which were made centuries before the Phoenicians began to make glass. Some scholars who study the history of ancient races – archaeologists, as they are called – are of the opinion that the method of making glass gradually developed from methods used to glaze pottery, for the glaze of the Egyptian pottery is very similar, chemically, to glass.[4] They also believe that this method, originated by the Egyptians, found its way to Phoenicia and other countries.

Where and when glass was first made is, therefore, something of an open question, but there is no doubt at all that the Egyptians were well skilled in making glassware by the reign of Emperor Tiberius, who ruled Rome during the lifetime of Jesus. During the preceding centuries the Egyptians had manufactured glass on a scale large enough to export it to foreign countries, and charged high prices for their glassware. Tiberius persuaded skilled Egyptian workers to set up glass factories in Rome and also to teach Roman workmen their methods. This venture was so successful that by the time of Emperor Nero (65 A.D.) Roman glass makers rivalled the Egyptians in making such things as mosaics and glass vessels. It is said, for example, that Nero ordered two glass drinking vessels at a cost of at least £50,000 of our money.

The story is told that during the reign of Tiberius a man discovered how to make glass which did not break into splinters when hit, as does ordinary glass, but merely became dented. He made a beautiful cup out of the glass, and, no doubt thinking of the great interest shown in glass making by Tiberius, decided to seek his favour by presenting the cup to him.

As with many old stories, there are a few different versions of the circumstances. One ancient writer stated that the man was a former architect who had been banished from Rome by Tiberius. The architect, in his place of retirement, discovered how to make unbreakable or malleable glass and then made a drinking vessel out of it. He hoped that on receiving such a novel and unique present Tiberius would not only pardon him and remove the order of banishment but would reward him handsomely.[5]

The historian Pliny gave only these few details of this wonderful discovery.

> It is said that in the time of Tiberius a combination was devised which furnished malleable or flexible glass; but that the workshop of the artist was

totally destroyed in order to prevent the value of copper, silver and gold from being depreciated. This report, however, was for a long time more widely spread than well authenticated.[6]

A fuller account was given by another writer, who related that

A Roman craftsman had discovered a way of making a glass cup that did not break, and, hoping to gain the Emperor's favour, decided to offer it to him. The vessel was greatly admired for it looked finer than any of the

The craftsman dashes his goblet to the ground

Emperor's golden cups. The man proudly passed the vessel to Tiberius but before there was time to inspect it closely took it back from the Emperor's hand and threw it on the ground. The Emperor was alarmed but the man picked the cup up from the ground and all the people then present then saw that it was dented, just as a bronze vessel would have been.

The man next took from his pocket a little hammer and with it easily and neatly repaired the damage by beating out the dents. This being done he thought that he was 'in Jove's heaven', particularly when the Emperor asked him, 'Does anyone but yourself know how to make this glass?' The man replied in the negative, whereupon Tiberius ordered him to be beheaded, because, if the secret became known, 'We should think no more of gold than dirt'.[7]

Some of these accounts were written by men who lived about the same time as Tiberius and they cannot, therefore, be totally discredited. Nevertheless, the fact remains that no 'unbreakable glass' appeared commercially for centuries. It has been suggested that the cup presented to the Emperor was made from a transparent resin which looked like glass but was not brittle and therefore did not break when struck.[8]

On a few occasions in the last two thousand years or so some one or other has claimed to have discovered an unbreakable glass. One such claim is of particular interest because the discoverer, like the one in Tiberius's reign, did not get the reward he had expected. This discovery is reputed to have happened in the time of the French King, Louis XIII. The inventor made a bust out of the new glass and offered it to Cardinal Richelieu, who was one of the most powerful statesmen France has ever had: in all but name, in fact, he was the real ruler of that country.

The Cardinal reacted rather as the Emperor Tiberius had done: and instead of giving the discoverer the reward which he expected, sentenced him to imprisonment for life because he feared that the glass manufacturers of France would suffer great loss of trade if a new, unbreakable glass became widely used. Hence the secret method of making this new glass, if it ever existed, was never disclosed.[9]

Nothing worthwhile resulted from any of these reputed discoveries. But one day early this century Edouard Benedictus, a French scientist, happened to witness a bad motor accident in which a woman was seriously injured by flying splinters of glass from the car window. This accident recalled to his mind a slight mishap which had happened to him a few years before with a substance called celluloid. Celluloid was then widely used for making knife-handles, combs, piano keys and various other articles, and was a cheap substitute for ivory and bone – today it has been almost entirely replaced by plastic materials. It is soluble in alcohol and a few other liquids, all of which readily evaporate.

In the year 1888, at the conclusion of an experiment with a solution of celluloid, a flask containing the solution was put on a high shelf in a laboratory. It remained there until one day in 1903 when Benedictus was tidying his laboratory. As he was taking the flask from the shelf it slipped from his hand and crashed to the floor. Benedictus expected to find a pile of broken glass and scattered splinters. To his great suprprise, however, he saw that although the flask was smashed yet the fragments were held together as if by some adhesive.

He picked up the wreckage and read the label which had been affixed to the flask fifteen years before. This informed him that the flask had contained a solution of celluloid. Benedictus realised that the liquid had completely evaporated during the fifteen years, leaving behind a layer of celluloid on the inner walls of the flask. He decided to keep the remains as a curiosity and attached to them a note stating the former contents of the flask and what had just happened.

When he witnessed the motor accident, Benedictus remembered the broken flask and, with a new born idea in mind, returned at once to his laboratory. It is said that he remained there all night and until well after day-break; by the end of that time he had thought of a method of making sheets of safety glass.

His method consisted of coating one side of a sheet of glass with a solution of celluloid and leaving it until most of the liquid had evaporated. Then, when the celluloid was tacky, he pressed another sheet of glass on top and left the 'sandwich' until the celluloid had set hard. The two pieces of glass were then firmly held together and when he shattered the sheet the splinters remained stuck to the film of celluloid. He had discovered a method that was to prevent injury from flying glass in many accidents.

There were three layers in this safety glass, two sheets of glass and a layer of celluloid, so Benedictus called his newly discovered sheets of glass by the name of 'Triplex', and patented the method of making it in the year 1909.[10]

* * * * *

There is little doubt that the smashing of the glass flask gave Benedictus the idea of making Triplex, but he was not the first man to take out a patent for a safety glass made in three layers. In 1906 an Englishman named John C. Wood had a similar idea, though he used a substance called Canada balsam where Benedictus used celluloid. Wood's invention, however, was not a commercial success whereas Benedictus' safety glass was soon in great demand.

Since 1909 many improvements have been made in the production of safety glass. In particular, new adhesives, and particularly those composed of plastics, have replaced celluloid.

2. *Hannibal Dissolves the Alps*

CARTHAGE, THE native city of the famous general Hannibal, was once a very proud city with a population of seven hundred thousand and an empire which extended along the northern coast of Africa and included also most of the islands of the Mediterranean Ocean and a settlement in Spain. During its short history the city struggled with Rome for supremacy in the Mediterranean regions but was finally defeated and utterly destroyed in the year 146 B.C. Its people were slain or dispersed, its buildings set on fire and ploughs were driven over its site.

Hannibal was born when the city was at the height of its power and was trained by his father, another Carthaginian general, in the arts of military warfare from a very early age. When he was only nine years old he went with the army to Spain and before his departure, on his father's bidding, he swore an oath of eternal enmity to Rome. This vow he kept all his life.

In 221 B.C. the soldiers in Spain proclaimed him ruler of the Carthaginian possessions in that country and he began to plan the fulfilment of his vow. All was ready for him to invade Italy and march on Rome in the spring of 218 B.C. and he set out with a force of 90,000 foot, 12,000 horse and 37 elephants. The elephants were to be employed in what today we should call shock tactics. Hannibal believed that when they were driven against the enemy the sight of the madly charging elephants would be so fearsome that it would throw the ranks of the soldiers into the utmost confusion.

Hannibal, instead of attacking by sea or the ordinary land routes, marched his army across the south of France and so to the foot of the Alps, a distance of one thousand five hundred miles.[1] His men, who were mostly natives of the warm lands of Africa, were greatly alarmed by their first sight of the snow-capped mountains which seemed to reach upwards to the very heavens. But Hannibal was not dismayed and ordered the ascent.

On their way up the mountain side they were attacked by the natives of the Lower Alps, who, despite the snow, ice and frost, lay in ambush waiting to kill any stragglers. On the ninth day of the ascent the army reached the top and halted, thoroughly weary and depressed. Hannibal saw that his men needed encouragement, so he called them together. The country was spread out below them like an open map and they could clearly see the wide, fertile plains of Piedmont. Hannibal pointed to the plains, saying, 'The way down is easy. Yonder is Italy. Yonder is Rome. After a battle or two all will be in our hands'.[2]

He was, however, greatly mistaken, for the difficulties to come were greater than those they had already undergone. The narrow track down the other side was so covered with snow and ice that often it could not even be seen and anyone who stepped off it fell to a terrible death down the mountain side. Soon the soldiers came to a very narrow part of the path which was completely blocked by fallen rock. The ground above and around was impassable: it was thickly covered by snow and ice. They could continue their march only by breaking a way through the barrier of fallen rock. Night fell and the army encamped.

Early next morning Hannibal ordered his men to fell the large trees which were growing near by, drag them to the fallen rock and pile them all around it. He waited for a strong wind and then

Vinegar dissolves the hot rocks

set fire to the wood. Soon the rock was very hot. He then ordered his men to pour vinegar on to it. The rock 'dissolved' and crumbled into pieces so that the soldiers were able to clear a way through it with their iron tools.[3]

The road down was now open. But the army was in a wretched state; the stores were failing; they had little fodder with them, and the rare grass on the mountains was covered with snow. However, Hannibal pressed on, and in time reached the Italian foothills of the Alps.

It had been a terrible march of fifteen days' duration. Two thousand years were to elapse before another general, Napoleon, attempted such a gruelling march; and he took a shorter and easier route. Hannibal's losses were tremendous; some thousands of his men perished amongst the mountains, and most of the provisions and baggage were lost. But he did not despair; after giving his troops a few days' rest he marched on to attack the enemy.

* * * * *

To scientists the most interesting part of this account of Hannibal's crossing of the Alps is the statement that he dissolved the rock with vinegar; and during the last two hundred years or so there have been some interesting discussions about this possibility.

An eighteenth-century chemist who made a close study of this crossing wrote that he would accept the story on two conditions: that it could be proved that Hannibal took vinegar with him on his march and that the rocks were limestone or marble.[4]

It is known that vinegar was one of the beverages of the Roman soldiers.* For example, Julius Caesar carried very

* A well-known occasion when weak vinegar was offered as a drink was at the crucifixion of Jesus. But the refreshing vinegar was deliberately spoilt by the addition of either gall, a bitter substance, or hyssop, which is also bitter. In St John, chapter 19, verse 29, the incident is thus described: 'Now there was set a vessel full of vinegar; and they filled a sponge with vinegar, and put it upon hyssop, and put it to His mouth'; and again in St Matthew, chapter 27, verse 34, thus: 'They gave Him vinegar to drink mixed with gall; and when He had tasted thereof, He would not drink.'

strong vinegar when on the march, diluting it with a very large amount of water before giving it to the tired soldiers. This refreshing drink was called *posca*. It seems reasonable to assume that the Carthagenians may well have had the same drink. Hence there is a likelihood that Hannibal had a supply of strong vinegar with him.

No one knows for certain the exact spot in the Alps where the rock fell but it might have happened in a region where the rock was largely of the kind which turns into quick lime when it is strongly heated; a kind which scientists call 'calcareous'.

Vinegar is a solution containing acetic acid which acts on calcareous rock forming the salt called calcium acetate; it also forms the same soluble salt with quick lime. Therefore, provided the rock was calcareous, Hannibal could have forced a passage through the landslide. But a very large volume of vinegar would have had to be used.

Livy was not alone in mentioning the use of vinegar for splitting rocks. Before his time Pliny had written that when cold vinegar is poured on hot rocks it will split them, even when attempts to split them by heat alone have failed. Vitruvius had also written that when rocks are heated in the fire and vinegar is poured on them 'they will fly asunder and are dissolved'.

The writer Polybius, however, made no mention of Hannibal's use of vinegar. This is a significant omission, for Polybius was not only the first to describe the march but he had made a special study of it and no doubt consulted many who were alive at the time. (He himself was a young child when the march took place.)

No other writer, it seems, mentioned this use of vinegar until Livy wrote his account as long as a hundred years after its reputed use. Other writers afterwards who have mentioned it may well have accepted Livy's story without verifying it.[5] Indeed at least one of them enlarged upon it in this way: 'Hannibal cut down the rock which was immensely high having previously dissolved it with a great force of burning flame and the addition of vinegar.'

The ancients knew of another way of breaking up big rocks.

The rock was strongly heated and cold water was thrown on it. This cracked the rock, then wedges, crowbars and other iron tools were used to widen the cracks until the stone was split apart.

There are many reasons, therefore, for doubting Livy's account of the use of vinegar. There is the uncertainty that the fallen rock was of the kind which turns into quick lime when heated strongly. Further, whilst it is very likely that Hannibal carried vinegar on the march there is a grave doubt whether the quantity he had with him, particularly towards the end of a long and difficult march, would be sufficient to deal effectively with a big fall of rock.

It is especially difficult to believe that Hannibal, or one of his officers, did not know that cold water and heat will split a rock. If they did then surely they would not have been so foolish as to waste vinegar when water in large quantities was readily available in the form of the snow and ice around them.

Then there is the lack of any mention of it until a hundred years after the march. Yet such a sensational use of the soldiers' drink would surely have been talked about in the taverns for weeks and weeks as soon as the men reached the populated parts. Indeed it would soon have become the subject of an 'old soldier's tale', possibly with some embellishments. Had this been the case it seems strange that Polybius, in his careful search for particulars about the march, never heard of it. Had he done so he would surely not have omitted to give a very full description of such an unusual event unless he had proved the story to be false.

Livy has not the reputation of being a careful historian, so the following suggestion is given for what it is worth.[6] It is said that a word formerly used in the northern parts of Italy for 'with an iron wedge' was *acuto*. This differs only slightly from *aceto*, the Italian word meaning 'by vinegar'. It may well have been, so the suggestion goes, that the soldiers broke down the rocks by driving iron wedges into them. The story would have been passed from mouth to mouth down the years and at some point the word *acuto* could have been replaced by *aceto*, possibly because some listener had not heard correctly. Then when Livy was gathering his material in readiness for writing his account, he could have heard the word *aceto* used in connection with this famous incident and not bothered to check his information from other sources.

3. Cleopatra Dissolves a Pearl

NOT ONLY was Cleopatra, Queen of Egypt, one of the loveliest queens of all time but she also had great charm, considerable intelligence and enormous wealth. These she never hesitated to use for her own purposes.

Forty years or so before the birth of Jesus, Mark Antony, one of the rulers of the Roman Empire, marched through Greece and Asia Minor compelling the people to obey the commands of Rome. During this expedition he learned that Cleopatra had been helping his enemies, so he demanded satisfaction from her. The Queen decided that she would visit him in order to answer in person the charges he had made against her. For she planned to use her charm, beauty and wealth to make him fall in love with her and so cease to be a danger to her.

The Queen sailed to the meeting in the royal galley, escorted by many smaller ships, in a river-procession of unsurpassed magnificence. The royal galley had sails of fine purple cloth, its stern was covered with gold and its oars were made of silver. The rowers kept time with the music from flutes, pipes and harps which sounded over the river. Under an awning of the finest workmanship, embroidered with gold, could be seen Cleopatra, exquisitely dressed to resemble Venus, the Goddess of Love, with boys dressed as Cupids fanning her with elaborately decorated fans. Beautiful girls dressed as sea nymphs controlled the sails with silken cords.[1]

Cleopatra decided to keep the advantage by remaining on board until Antony, out of curiosity, went to see the magnificent spectacle. In anticipation of his visit braziers of incense had been lighted and soon the perfume was wafted to the crowds on the banks of the river. As twilight fell fairy lights were set high on the masts in a variety of figures and made a most impressive sight.[2]

Antony boarded the galley to question the Queen but he was soon won over by the charms of the beautiful, designing woman and agreed to stay on board to dine with her. Careful preparation had been made for him.

The floor of the dining cabin was covered with a layer of flowers many inches deep, whilst the couches and the walls were covered with embroideries in purple and gold. The food was served from dishes of gold inlaid with precious and glittering stones; the golden goblets from which they drank were richly ornamented with gems; the food was rare and costly. Antony, completely overwhelmed by all he saw, was loud in his praise.

Cleopatra was too clever to acknowledge that she had made special arrangements for him and set out to give him the impression that she normally lived in this manner. Indeed, to try to show that there was nothing unusual in this way of living, she made him a present of everything used at the banquet – the couches, the golden dishes, the gem-encrusted goblets – everything. She then invited him to remain to a party on board. This he did and they spent some pleasant hours together, dancing and revelling.

Other parties of a similar extravagance followed and Antony was deeply impressed, commenting on the huge sums which the banquets must be costing. Cleopatra replied that the cost seemed trifling to her and told Antony that if he wished to attend what she would consider a really extravagant banquet she would arrange one costing about ten thousand sestertias (probably about £200,000 of our money today).

Antony replied that it was impossible to spend so much on one banquet, whereupon Cleopatra laid him a wager that she would do so on the following day. The wager was accepted, and Plancus, one of his generals, was appointed 'judge of the wager'.

Next day Antony and his generals again went on board. At

first the banquet seemed no more costly than any of the others, but towards the end of it Cleopatra announced that the cost so far had been negligible and that she now proposed to spend ten thousand sestertias on herself alone. She was lavishly adorned with gems and jewels, and from each ear dangled a huge pearl.

Cleopatra and the pearl

She called for a goblet of vinegar; this her attendants brought and placed before her. Quickly she took a pearl from her ear and dropped it into the vinegar, and, whilst everyone gasped with amazement, drank the liquid at one gulp. She then made to take the pearl from the other ear. But Plancus, the judge of the wager, very quickly stopped her by announcing that she had already won her bet.

3. Cleopatra Dissolves a Pearl

This story of the pearl and vinegar was told by Pliny and is usually accepted as true. Other similar incidents have been recorded.[3] For example, 'a man about town', a Roman named Clodius, inherited a large sum from his father (Aesopus the writer). He boasted that he would dissolve a valuable pearl in vinegar not just to win a bet as Cleopatra had done but to discover how pearls tasted. This he did, and because he found the drink 'wonderfully relishing' he gave a pearl to each of his guests to do the same.[4]

* * * * *

Pliny, who described the incident of Cleopatra and the pearl, also gave many recipes for medicines, one of which, a cure for gout, was made by dissolving small and almost worthless pearls in vinegar, but they were added only after they had been ground to a fine powder.

Powder obtained from pearls is composed mainly of calcium carbonate, which is soluble in all acids, including vinegar; it also contains a little insoluble ash. But a whole pearl is protected by a skin, which does not dissolve in a few seconds in vinegar of a strength weak enough to be drunk without causing injury. It is therefore highly improbable that Cleopatra's pearl would have dissolved quickly enough for her purpose had she put it in drinkable vinegar.

A few ingenious suggestions have been made to account for the incident. One is that Cleopatra, who was well skilled in the chemistry of her day,[5] might have added a substance to the vinegar before the banquet which would have dissolved the pearl. But the author of this suggestion does not name the substance. Another suggestion is that Cleopatra made a dummy pearl of chalk and wore it purposely so that she could pretend to dissolve a real pearl in vinegar. But such an act would not be in keeping with her character.

Another possibility is that she put a real pearl into the cup and swallowed it whole, along with the vinegar, pretending that it had actually dissolved.

The story of Cleopatra and the pearls has been mentioned by so many classical authors that it is difficult to believe that it is wholly fictitious. Indeed, one author gave the subsequent history of the remaining pearl. According to him, it was taken to Rome where it was cut in two and used as ear-decorations on a statue of Venus.[6]

* * * * *

A somewhat similar story of the dissolving of pearls in vinegar or wine is told about the fabulously rich Elizabethan nobleman, Sir Thomas Gresham.[7] In 1564 Sir Thomas built a large building, so that the merchants of London could do their business in comfort instead of having to transact it walking and talking in an open narrow street, where, according to a historian, they had 'either to endure all extremities of weather or else to shelter themselves in shops'.

Sir Thomas and the pearl

The building, a magnificent edifice, was opened by Queen Elizabeth in 1571. Accompanied by her nobility and courtiers, she dined in state with Sir Thomas. The dinner was in keeping with his great wealth and extravagance. But the royal toast at the end of the meal was the most extravagant of all. For Sir Thomas laid on the table a most magnificent pearl, crushed it into a powder, put it in his wine, and standing up, drank Her Majesty's health.

The Queen, with Sir Thomas and the courtiers, then inspected the new building. Her Majesty, we are told, 'having viewed every part thereof, caused the name, by herald and trumpet, to be proclaimed the Royal Exchange and so to be called hence-forward and not otherwise'.

There is no record in any account of the opening that this incident actually occurred and no mention is made of it in the authentic histories of that period. Indeed, the only mention appears to be in a play which included the banquet scene, and mentions the incident of the pearl in these words:

> Here fifteen hundred pounds at one clap goes,
> Instead of sugar Gresham drank the pearl
> Unto his Queen and Mistress; pledge it Lords.

4. The Monk and Gunpowder

BERTOLD SCHWARZ was a monk of the Franciscan order who lived during the fourteenth century at either Nuremberg or Freiberg in Germany. Very little reliable information is known about him; indeed, historians are not at all sure about his actual name. Some refer to him as Constant Aucklitzen and others as Niger Berchtoldus. Frequently he is named simply Black Bertold. (This is quite a good name for him, because in his day science was looked on as black magic by many people; moreover the name chemistry probably comes from a word meaning 'the black art').

Bertold, who was accustomed to making medicines for the peasants living near the monastery, was making up a mixture for one of his patients which contained sulphur, nitre and charcoal.[1] He had probably ground each substance in turn into a powder in his mortar, and then mixed them carefully together. He left the mixture in the mortar, covering it with a large round stone. Some time later he wanted a light; so he struck his flint. A few sparks, by chance, flew into the mortar and set fire to the mixture. Immediately there was a loud bang and the stone was hurled upwards with so much force that it went straight through the roof of the building. When Bertold had recovered from the shock he saw his empty mortar, and, in the roof above it, a round hole through which the stone had passed.[2]

The stone is hurled upwards

A German writing in 1743 enlarged on this story. According to him Bertold was eager to study fully the properties of this wonderful substance which could hurl a stone with so much force. So he made another supply of it. Then, so the German wrote, 'Being curious to know the force of his powder he was so silly as to fill a leather bag with it, then to put his feet on the bag and then to set fire to it by a long train of powder'. The result of this rash experiment 'was that he was forthwith blown up and his brains were dashed out against the ceiling of his little cell'.[3] This part of the story is almost certainly fictitious; nobody who had seen a stone hurled with great force through a thick stone roof would be likely to perform such a silly experiment.

The accidental projection of the stone through the roof of the monastery gave Bertold the idea of using gunpowder to hurl stones in battle. It is believed that in his first attempts he used either a mortar or else something of a similar shape but a little longer. In the bottom of this he placed a mixture of gunpowder and put a large stone on top. Then he set fire to the gunpowder and the stone was hurled out. But even when the mortar was pointed in the desired direction such a 'firearm' would not have a good aim. Therefore it is probable that only a short time elapsed before a long iron tube replaced the stumpy mortar. The tube would, of course, be closed at the end containing the powder, but a small opening would be left through which the explosive could be lighted.

In those days the process of casting iron was unknown and it is thought that the tubes were made of strong iron bars put together like the wooden staves of a barrel and strengthened with hoops as a barrel is. At a later date the tubes, or cannons as they were called, were cast in one piece in brass or iron.

There is much evidence to support the belief that Bertold invented such a cannon, though the invention may not have been due to the accidental explosion in a mortar.[4] But even that possibility cannot be ruled out completely, because for centuries the name mortar has been given to a piece of ordnance whose shape might have been evolved from that of the mortar used by chemists. Its barrel was short and stubby with a very wide bore and it fired its shot at a very high angle. The illustration, which is based on a medieval print, shows four mortars hurling their shot over the high walls of a fort.

Medieval mortars in action

An authority on artillery, named Guttmann, quotes a regulation of the French Mint dated 17 May 1354 which states that the King of France, after having satisfied himself that artillery was invented in Germany by a monk named Bertold Schwarz, ordered the General of the Mint to make inquiries about the metal required to make cannon.[5]

Another historian, whilst stating that Bertold lived thirty or forty years after the gun was first discovered, admits the possibility that Bertold might have been a 'specialist in artillery'.[6]

The introduction of gunpowder in warfare brought about many changes in the methods of fighting, particularly when hand firearms were introduced about 1500. The catapults and battering rams of previous wars were replaced by much more powerful weapons. One of the most famous of all early guns was that used by Mahomet II at the siege of Constantinople in 1453. Tradition has it that it could throw a stone weighing over six hundred pounds for hundreds of yards with such force that when the stone hit the ground it buried itself 'a fathom deep'. To drag the gun along thirty waggons were linked together and were drawn by a team of sixty oxen; two hundred men on each side were stationed to prevent the gun from toppling over sideways, and two hundred and fifty workmen marched before it to smooth the way and repair the bridges. With this and other guns, Mahomet, 'the first gunner' in the world, quickly captured the City of Constantinople, despite the three thick walls which had been built around it as a protection against a siege.

After the introduction of gunpowder and artillery the influence of the great war-loving nobles gradually waned. Guns and gunpowder were very expensive and few individual lords could then afford to keep their own 'private army'. In Britain and in many other countries the king of the realm, and later parliament, being financed by taxes and the revenues of the state, became the masters of the armed forces.

There is also little doubt that the use of gunpowder greatly helped the 'civilised' nations in their wars against the native races. For example, the very rapid conquest of South America in the sixteenth century was largely due to the use of guns and gunpowder by the invading Spaniards, for against these weapons the bows and arrows and poison darts of the natives were of little avail (see chapter 7).

It is interesting to note that complaints were made about the first use of gunpowder: we read, 'All Italy made complaints against the use of gunpowder as a manifest contravention of fair warfare'; and the knights of old loudly protested against 'the villainous saltpetre' and against the new 'unknightly way of fighting'.

A famous author, writing about the year 1500, summed up the thoughts of many men of his day in these words:

> But of all other things that were devised to the destruction of men, the gun be the most devilish, which was invented by a German whose name is not known... For this invention he received this benefit,– that his name was never known, lest he might, for this abominable device, have been cursed and evil spoken of as long as the world standeth.[7]

Thus in the sixteenth century too, the introduction of a new weapon of destruction caused alarm, bitterness and complaint just as it did in 1915, when poison gas was first used, and in 1945 and later years when the atomic bomb was introduced (chapters 21 and 41).

5. *How Antimony got its Name*

ANTIMONY IS a white silvery-looking metal of fairly recent discovery, but its compounds, and especially that with sulphur, were known to the ancient peoples; the frequent references in the early histories to the word antimony are not to the metal itself, but generally to its sulphide.

The sulphide was employed, according to a well-known historian of chemistry,

> by the Asiatic ladies in painting their eyelashes, or rather the insides of their eyelashes, black. Thus, it is said of Jezebel, that when Jehu came to Jezreel she painted her face. She put her eyes in sulphuret of antimony. So also did the woman in Ezekiel. This custom of painting the eyes black with antimony was transferred from Asia to Greece, and while the Moors occupied Spain it was employed by the Spanish ladies also.[1]

One of the fascinating stories in the long history of the metal gives the reason why its name was chosen and also tells of one of the most famous of all the alchemists – real or legendary.[2] It relates that this fifteenth-century alchemist, Basil Valentine by name, lived in a monastery at Erfurt, a town in Saxony, along with his brother monks of the learned Benedictine order.

In medieval times it was the custom for alchemists to conceal their true names by adopting fanciful ones. The name by which this monk had called himself had a most distinguished meaning. For the Christian name of Basil is equivalent to a Greek word meaning 'the King', whilst the surname Valentine comes from a word *valentino*, meaning 'the mighty'. So his full name meant 'the mighty king' (that is, of course, the mighty king of the alchemists).

Basil Valentine appears to have been a distinguished alchemist and his writings gave a thorough summary of the chemical knowledge of his day. A romantic legend relates how, shortly before his death, he put his manuscripts under a table of marble behind the High Altar in the Cathedral church of Erfurt deliberately hiding them in the belief that they would be miraculously brought to light whenever the time came for their revelation. Many years later the 'time' came; the cathedral was struck by a thunderbolt and the walls were knocked down, revealing the manuscripts!

Valentine was led to study medicine because he wanted to find a cure for a sick monk in the monastery. For this purpose he tried to find a suitable herb: although he was not successful in his search, his study of herbs led him to become an enthusiastic alchemist. He had no laboratory and so did all his experiments in his own small cell.

The monastery at Erfurt, like other Benedictine monasteries, housed a self-supporting community, with its own farm and live-stock. It was the custom of those days to allow the animals and fowls to roam about the monastery grounds almost wherever they wished, foraging for food. Part of their food consisted of the scraps people threw away, for people of his day did not use dustbins or waste boxes but threw their litter out of the door or window into the street or on to the spare land about the building. The litter was left to lie where it fell as nobody was at all concerned about it.

So when Valentine finished his experiments he used to throw out of the window the substances which he had finished with – the residues, as they will now be called in this story. These residues fell on a pile of rubbish which had gradually accumulated beneath his window.

One day the monk chanced to see the monastery pigs busily rooting in the pile and thoroughly enjoying the residues. He wondered what the effect would be, and decided to keep them under observation for some time, expecting that the animals would be unwell after their very unusual food.

To his surprise he found that the residues had not harmed the pigs in the least, but, on the contrary, seemed to have done them good. For whereas formerly the pigs had been lean and thin they were now growing fat and strong.

Valentine had often noticed that some of his fellow monks

Basil and the pigs

looked thin, weary, and ill nourished, and his experience with the pigs led him to believe that the residues would do the sickly looking monks a world of good. So he persuaded them to eat a portion of the residues.

Unhappily the new 'medicine' was much too drastic for some of them, for they were really in poor health and not as fit as the pigs had been before their dose. Consequently these unhealthy monks could not stand the shock to their system and many of them died.

The poisoning of his fellow monks greatly distressed Basil, and to prevent a like occurrence in future he gave the residues a name which would warn every one of their poisonous nature. This name was antimony; and he chose it because *anti* means against and *moine* means a monk.[3]

After this terrible tragedy Basil made a careful study of antimony and discovered that a small dose of it was a very good medicine. (It should be mentioned that in those days the word antimony was used for any preparation containing the metal.)

This romantic account of the origin of the name is now discredited for it is known that the word was used as early as the eleventh century. Moreover, the story implies that the derivation of the word comes from two words, *anti* and *moine*, the latter of which is a French word, whereas Basil was a German![4]

Probably one reason for the persistence of the story is a book written by Basil which contains a passage that could easily have started the story, thus:

> If a man wishes to fatten a pig, let him give to the pig, two or three days before he begins to fatten it, half a drachm of crude Antimony, so as to cause a thorough cleaning out of the bowels; the pig will then eat more freely, fatten more quickly, and be freed from any bilious or leprous disease to which he may be subject. I do not advocate that crude Antimony should be administered to human beings: the brutes can digest raw flesh, and many other things which would overpower the strength of the human stomach.[5]

It will be noted that he specially warned anyone against the use of much 'crude antimony' on human beings, possibly because he knew by experience the result of doing so.

Not only has this particular incident in the life of Valentine been challenged but many historians now doubt the existence of Valentine himself. One of them wrote:

> His works were spread abroad by means of copies and excited the interest of the Emperor Maximilian I to such a degree that he caused a searching inquiry to be made in the year 1515 as to which Benedictine convent the famous author had dwelt in; but unfortunately his efforts in this direction were without result, as were also all later ones.[6]

Other writers have stated that the works reputed to have been written by Valentine are undoubted forgeries containing some facts which were not discovered until a hundred years or more after the alleged date of Valentine's death.

6. *Alum, Popes and Kings*

ALUM HAS been known for at least five hundred years. It is obtained from a particular sort of stone, known as alum-stone, which occurs just below the soil in certain parts of the world. It is manufactured by a fairly simple process and the stone is readily mined since it occurs so near the surface. Alum has had many uses; its main one, however, was in the dyeing process, for it gave a brightness to many colours, which were naturally dull, and fixed some of the medieval dyes firmly to the cloth so that they were not washed away.

In the fifteenth century alum was a very valuable substance and most of it used in Europe was obtained from the neighbourhood of Constantinople, where there were very rich mines. When the Turks captured that city in the year 1453 they immediately confiscated the mines and thereby became the chief manufacturers of alum in the then known world.

Before the fall of the city an Italian named Castro had been living in Constantinople, trading in cloth and dyestuffs, and in the course of his work had gained some knowledge of alum. In 1453 he succeeded in making his escape and returned to his native land. Many years later, when walking in the hills in the district near Tolfa, he happened to notice that the grass had the same colour as that growing near the alum-stone mines in Constantinople. He picked up a few of the white stones and bit them. They had the same salty taste as the stones near the alum mines. After making other tests he was sure that he had found a supply of the valuable alum-stone and at once left Tolfa to tell the Pope of his wonderful discovery in these words:

I announce to you a victory over the Turk. He draws regularly from Christians over three hundred thousand pieces of gold, paid to him for alum. I have now found seven hills which are so abundant in alum that there is almost sufficient of it to supply seven worlds. It you will send workmen and cause furnaces to be built you may supply alum almost to all of Europe. Wood and water are both in plenty there, and you have a port where vessels bound for the west may be loaded. You can now make war against the Turk; this mineral will supply you with the sinews of war, that is money, and at the same time will deprive the Turk of them.[1]

The Pope thought that these words of Castro 'were the ravings of a madman, and all the cardinals were of the same opinion'. But though his proposals were often rejected, Castro did not abandon his pleadings, and finally persuaded the Pope to send skilled men who had worked in the mines in Constantinople to investigate his find. They, after having closely examined the ground, declared it was similar in appearance to that of the Asiatic mountains which produced alum. 'Shedding tears of joy', we read, 'they kneeled down three times, worshipping God and praising His kindness in conferring such a valuable gift. The stones were baked and produced alum more beautiful than that of Asia, and superior in quality.'*

The Pope lost no time in building his alum works. The stone was first baked in a furnace (as shown at the top of the illustration) and then thrown into water (as in the centre of the picture). The alum in the stone dissolved whilst the impurities fell to the bottom of the vessel. The solution was then put into a large leaden vessel (shown in the bottom left corner of the picture) where it was heated until it would crystallise. Finally it was run into a wooden vessel where it stood until all the crystals had come out of solution. The result was 'alum of the most perfect kind'. We are told that the Pope, Pius II, 'sensible of the great benefit which might arise to the Apostolic Chamber, employed more than eight hundred people at Tolfa in preparing it'.

In a few years the Pope, now Julius II, was receiving a huge income annually from these alum works. He announced his intention of using these funds to wage war against the Turk and therefore declared that the sole right of manufacturing alum was reserved to himself and that anyone else making it would be

* The alum-stone of Tolfa is of volcanic origin and is impure hydrated potassium sulphate mixed with iron oxide, alumina and clay, as impurities. When it was roasted it became dehydrated and when the product was thrown into water, the sulphate dissolved and the impurities settled to the bottom of the vessel. The solution was then evaporated to crystallisation point and finally run into the wooden vessels where it was allowed to stand until cubic crystals of alum, $K_2SO_4 . Al_2(SO_4)_3 . 24 H_2O$, came out of the solution.

committing an offence. The Pope also made it an offence to purchase alum from the Turks. The penalty for either offence was excommunication, a penalty greatly dreaded by all devout Catholics.

The medieval manufacture of alum

Protestants, however, cared nothing about the wrath of the Pope or about excommunication, and a romantic story tells how an English protestant defied him and set up alum works in this country.[2] The story is similar, but only at the outset, to that of Castro's discovery on the hillside at Tolfa.

This Englishman was 'the learned naturalist Sir Thomas Chaloner'. He

observed that the leaves of some of the trees growing near Guisborough (which is in Yorkshire) were of a peculiar shade of green. The oak roots were very broad but did go deep into the ground; they had much strength in them but little sap. The soil was a white clay sprinkled with yellowish and blue colours, and never froze. On a pretty clear night it shone like glass.[3]

Sir Thomas had seen similar sights in the alum-bearing district of Italy and began to wonder whether he had discovered an alum mine near Guisborough. He had tests made, and to his great delight they revealed a plentiful supply of this very valuable alum-stone. He decided to set up an alum works in Guisborough.

But it was hard for him to discover the methods used by the Pope in the manufacture of this substance. Not only were the papal works carefully guarded from all strangers, but anyone found trying to discover the secret method was put to death. Sir Thomas, at the risk of being caught, went to Italy and bribed two or three of the men working in the alum mines to return with him to England.[4] It is said that he hid them in large barrels which were labelled 'for England'. The port officials had no reason to suspect that the barrels contained anything unusual and allowed them to be put on board the next British sailing ship which called at the port.

On arriving at Guisborough the Italians established an alum works and taught the local men the new job. The Pope, on learning this, thundered out against the knight and the fugitive Italians the most complete curse known. It was the one first uttered by the monk Ernulphus a few centuries earlier. In it, every part of the body, from head to foot, is cursed in the most dreadful language.

Chaloner, however, did not worry about this and the new

The workmen hide in the barrels

business flourished, especially when the Sovereign gave the Chaloners the sole right of manufacturing alum in this country. Hence all went well until another King, Charles I, needing money, decided to take possession of this most valuable alum mine.[5] He did this simply by declaring that the mine was a mine-royal, for all profits of mines-royal went into the royal coffers.

Naturally the Chaloners were angry at losing this valuable source of income but they were powerless to do much about it at the time. In 1641, however, when Parliament revolted against the King and bitter fighting took place, they saw a chance of getting their revenge by joining the side of Parliament.

As is well known, Charles was defeated and, in the year 1649, was put on trial 'as a tyrant, traitor, murderer and a public enemy of the Commonwealth of England'. The trial of the King was held at Westminster before one hundred and thirty-five men who had been appointed commissioners or judges. Charles refused to acknowledge their right to try him and would not plead; he was sentenced to death. The death warrant authorising the execution was signed by only fifty-nine of the judges. One of these was 'Tho. Chaloner'.

But the traditional story does not end with Charles's death. It tells how Thomas Chaloner was made a Governor of the Isle of Man by Cromwell where he lived in one of the castles, well attended and in great state, until the year 1660. In that year the Stuart line was restored and Charles II was crowned King. He had agreed to forgive everyone who had rebelled against his father, except the judges at the trial.

Chaloner, in his castle in the Isle of Man, heard the long-dreaded news that His Majesty's soldiers were coming to arrest him.[6] He called for his servant girl to bring him a drink, and put into it a poison which he had been carrying in expectation of such a time. He drank. When Charles's men entered the castle they found him dead.

* * * * *

The stories about alum are a mixture of truth and romance. The account of its discovery in Italy was recorded by Pius II, the Pope at the time it happened; it is, therefore, accepted as true. The account of Chaloner's accidental discovery of alum stone is so very similar to that of Castro's discovery that there is some justification in believing that the details of the first find were transferred to the description of the second.

It is known, however, that a Chaloner did start the alum works at Guisborough, either in the reign of Elizabeth or in that of James I; and it is also known that these did become crown property in Charles I's reign.

An historian of Whitby (which is near Guisborough) dismisses the rest of the story 'as strangely erroneous both as to time and person'.[7] He points out that it confuses three Chaloners. First there was Sir Thomas who founded the works; then came Thomas, his son, who was an eminent scholar, a member of the Long Parliament, and one of the King's judges. It is his signature that is on the death warrant of Charles Stuart. Then there was a brother James, who was also a member of parliament and one of the judges. It is a fact, however, that the family did side with Parliament and the reason may well have been that they had been treated unfairly over the alum works.

At the Restoration in 1660 Sir Thomas was dead; his son Thomas fled to Holland, where he later ended his days, but James was living in the Isle of Man as the story relates, and was actually summoned to London. Fearing the consequences of the leading part he had played in the King's execution he poisoned himself, as the story tells.

It is therefore probably true to say that the latter part of this story of alum is a legendary tale made up of episodes from the lives of Sir Thomas Chaloner and his two sons, Thomas and James.

7. *Gunpowder and a Volcano*

AFTER THE first voyage of Columbus to the West only a few years elapsed before the Spaniards began to establish settlements on the mainland of the American continent, attempting, in the year 1518, to found one in the country now known as Mexico.

The Spaniards discovered, to their great surprise, that the natives of Mexico were a much more highly organised people than the rough and uncouth islanders whom they had previously encountered and whom they regarded as savages. For these natives of the mainland had a well organised mode of life with a properly constituted form of government, a sound legal system, and a monarch who lived in a stone palace, in great pomp, with courtiers and ambassadors in attendance. Moreover, the common people were skilled in many occupations, and worked in copper, tin, gold and silver.

7. *Gunpowder and a Volcano*

The king was an experienced warrior, and his soldiers were fierce and fearless, since they believed that any man who fell in battle was immediately transported to a region of ineffable bliss in the bright mansions of the Sun. Their generals wore gold and silver ornaments and were clothed in tunics made of a quilt-like material, two inches thick, which gave much protection against flying arrows. The common soldiers daubed themselves with war paint before going into battle and fought sometimes with arrows tipped with stone or bone, sometimes with wooden spears which had heads of bronze and sometimes with wooden clubs with heads on which were two rows of flaked stones with sharp points or edges.

In 1518 a young Spanish general named Cortes was sent by Velasquez, the Spanish Governor of Cuba, to conquer Mexico. He had only a small force of between six and seven hundred soldiers with eighteen horses and some pieces of cannon. Such a small force was considered large enough to deal with thousands of savages, and there was no reason to suspect the existence of an organised race of people.

Cortes set sail in November 1518 with a fleet of seven vessels, and, after many preliminary adventures, built a port on the mainland which he named Vera Cruz. Then, to make retreat impossible, he destroyed all his ships.[1] From that time onwards his men knew that they must conquer or perish.

The natives, despite their disregard of death, could not battle for long against cannon and shot, and in 1521 their capital, the city named Mexico, fell to Cortes. Soon the whole country submitted to his rule.

The conqueror at once ordered the building of a new Mexico on the ruins of the old city and decided to fortify it. But affairs took an unfortunate turn for him. He found himself in serious want of artillery and ammunition. The Governor, Velasquez, had become not only envious of his success but bitterly hostile to him. Hostile also was the chief of the Spanish Colonial Department which governed these expeditions, and these two powerful men schemed in such a way that Cortes was kept short of many things, and in particular of artillery and gunpowder.

Cortes burns his boats

Cortes, although dismayed, was determined to succeed. He decided that if he could not get artillery and gunpowder from Spain he must get them in Mexico. But this task was a tremendous one. The manufacture of gunpowder in the New World had not been thought of, whilst iron, out of which the Europeans made not only their cannon but also some of their cannon balls, had not then even been found in Mexico.

Cortes did not despair. If he could not use iron to make cannon he knew he could use bronze. For he had learnt that the natives had some articles made of a bronze which was composed of copper and tin.[2] There was a plentiful supply of copper in Mexico. As for the tin, he knew it must occur naturally in that country, since the natives used it not only to make their small supplies of bronze, but also as money: (their coins were made of sheet-tin cut into the form of a capital T). A search revealed a plentiful supply of this metal in a district called Tasco, and there he opened tin mines and built a foundry. The copper and tin were melted together to form bronze and the molten bronze was cast into moulds. In all, thirty pieces of cannon were made in this foundry, and, taking into account the cannon he already possessed, he was satisfied that he was well supplied with pieces of artillery.

The supply of cannon balls did not cause him great concern. Although the Spaniards generally used iron cannon balls, they, like the gunners of other countries of that time, did occasionally use balls of stone. Obviously, in the absence of iron they could use stone; and therefore he ordered that a large number of cannon stones should be made from the crude stone of the region. Now all that he needed was gunpowder.

Gunpowder is a mixture of charcoal, sulphur and a white substance known as nitre or saltpetre, which is very rich in oxygen. When the powder is set on fire the charcoal and the sulphur burn in the oxygen given to them by the nitre and give off an enormous volume of gases almost instantaneously. This large volume of gas, when produced in the confined space of the barrel of a cannon, exerts a very powerful force which ejects the shot from the cannon's mouth.

The method of making charcoal which was employed in those days was known either to Cortes himself or to one of his men. Three long pieces of thick wood were made into a triangle and

Cortes makes charcoal

put flat on the ground in a clear space in a forest, and a long pole was thrust upright into the ground at the centre of the triangle. Logs of wood, from which all the small branches had been cut, were put on the triangle so as to form a pile around the pole. The pile was then covered all over with earth.

The wood was set alight round the bottom of the pile. Some of the logs burned and set fire to the others but, because the covering of earth prevented air from circulating freely through the pile, most of the wood did not burn away but merely smouldered to form charcoal.

Cortes therefore, in this way, readily obtained a supply of charcoal.

Nitre or saltpetre occurs naturally in warm countries. It is found either as a thin layer on top of the soil or mixed with the surface layer of soil. When the mixture of nitre and soil is put into water, a solution of saltpetre is formed from which crystals can easily be obtained. There was a good supply of saltpetre in the soil as well as in many caves near Mexico city.

The way Cortes obtained his sulphur makes a very interesting story. A few years before his shortage of gunpowder, he and his men had on one of their marches passed close to a very high mountain named by the Indians Popocatepetl.[3] This name means the 'hill that smokes', for the mountain is a volcano. The natives regarded it with awe and fear and many legends had grown up about it. Some believed it to be the abode of the departed spirits of wicked rulers, who bellowed aloud in their agony; others thought it was the dwelling place of the Gods. It is not surprising that the natives dared not ascend this mysterious and terrifying mountain.

When Cortes first approached the volcano in 1519, it was in a state of activity, raging with uncommon fury and throwing out fire, ashes and smoke.

The Spaniards knew that the natives did not dare to climb the mountain. One of their captains and two of his men therefore determined to do so, to show the natives that Spaniards were members of a superior race. A few of the natives were persuaded to set out with them but they would go only as far as a certain spot on the mountain side. This was the place which tradition marked as the entrance to the dwelling place of the Gods.

The Spaniards, to the awe of the natives, entered the abode of the Gods and climbed higher up the mountain, but they, too, turned back before reaching its summit. For further progress upwards was impossible, the top being covered with snow and ice. But this was not the worst difficulty, for the volcano was belching out ashes, as well as clouds of smoke which smelled of sulphur.

They returned, taking back with them only snow and pieces of ice, the appearance of which, according to Cortes, 'astonished us very much because this country being near the equator ought to be very warm'.[4] (Cortes had not realised that even on the Equator the temperature at very high altitudes is below freezing point).

In 1521 Cortes recalled that this expedition had encountered sulphurous smoke and reasoned that there must be sulphur on this mountain. He realised that it would be a dangerous job to get any from such a place. Nevertheless he knew also that he had some courageous men in his army. So, at this time of desperate shortage of the gunpowder on which his power depended, he did not hesitate to send up the mountain a picked party of five men under the leadership of Francisco Montano.[5]

The volcano was quiet at the time, so on this occasion the party had not to face a storm of ashes and smoke. After a difficult climb they reached the top and made their way to the very edge of the crater. Before them yawned a large gulf; it had an irregularly shaped mouth which was between one and two thousand feet wide. They peered downwards and estimated that the depth was between eight hundred and a thousand feet. A lurid flame was burning at the bottom of the abyss and sending up 'sulphurous steam' which, cooling as it rose, left solid sulphur on the rocky walls of the cavity. They had hoped to find sulphur around the rim of the crater but were disappointed. There was, however, a good supply in the depths of the crater. They decided to fetch some of it.

The descent into the crater would obviously be a terrible task and they therefore agreed to cast lots to settle which one of them should descend. The lot fell to the leader himself. They had taken baskets with them to transport the sulphur back to Cortes: one of these was dragged to the crater's edge and then gently lowered, with their leader in it, down the crater until a depth of about four

Montana descends into the volcano for sulphur

hundred feet was reached. There Montana hastily scraped enough sulphur from the rocky walls to fill the basket, and was hauled up. Again the descent was made and yet again, until he had made seven trips down the crater. At last, when about three hundred pounds of sulphur had been gathered, the Spaniards made their way down the mountain side, back to the city of Mexico. They had collected enough sulphur to make fifty kegs of gunpowder, a supply which was adequate for a considerable time.

Such, then, were the resources displayed by Cortes. They enabled him to supply every deficiency and to triumph over every obstacle which the malice of his enemies had put in his path. Cortes, in reporting to his King how he had obtained his gunpowder, wrote somewhat ironically that 'it would be less inconvenient, on the whole, to import the powder from Spain'.[6]

8. Epsom Salt

THE SUMMER of the year 1618 was a very dry one and water was scarce in many places on the downs in Surrey. This scarcity was keenly felt in the little village of Epsom, and the cattle were in dire straits through thirst. Henry Wicker, like many of the farmers of the district, had very little water for his cows.[1] One day, whilst walking in one of his fields, he came upon a small hole in the ground. To his great surprise, for rain had not fallen

The cows refuse to drink

for days, this hole was full of water. He reasoned that there must be a spring of water near by; so he sent for his men to dig up the ground round the hole. A spring was found and it yielded a steady flow of water. His water problem seemed to have been solved, and his men dug a hole large enough for a pond where he could water his cattle.

He sent for the beasts and put them in the field containing the pond. The cows eagerly went to it, but, to Wicker's dismay, none of them would drink. Realising that there must be something unusual in the water, he sent a sample to the analyst and received a report that the water contained alum, a substance with a bitter taste. Alum was, however, much used then in dyeing and for cleaning and healing sores, cuts and wounds (see chapter 6), so Wicker had a profitable use for the pond-water.

Another lucky incident happened one day during the summer of 1630 when some workmen near the pond became very thirsty, and seeing the water in it, quenched their thirst, despite the bitter taste. It soon became apparent that the water had a purging effect and obvously contained a substance other than alum. So another analysis was made. This showed that the water contained the substance now known as magnesium sulphate.

Another version of the traditional story puts the date earlier than 1618 and reads:

> Towards the end of the reign of Queen Elizabeth, notice was taken by some persons, that the water of a pond on the common, half a mile west of the village, had performed great cures on many country people, who were troubled with ulcers and other disorders.
>
> In the reign of James I some physicians heard of the water's fame and visited Epsom. They analysed the water and, finding that it contained a 'bitter purgative salt', reported that it was the first pond of its kind to be found in England. These physicians spread its fame far and wide, and in time so many people visited Epsom to take its waters that the Lord of the Manor decided to enclose the pond or wells with a wall and erect a shed (or rest house) for the sickly visitors.

An account written in the seventeenth century mentions many of the distinguished visitors who went to Epsom to drink 'a medicine sent from heaven'. One of them was Maria de Medici, mother of the wife of Charles I, and her example was followed by numerous 'society folk' and others in Stuart times. After the Restoration, Nell Gwyn kept 'merry house' there and so Charles II was a frequent visitor.[2]

By the end of the seventeenth century Epsom had become the centre of fashion. A large and pretentious building with a ball-room seventy feet long had been built; taverns reputed to be the largest in England had been opened and the streets were thronged with people in carriages and sedan chairs, or on foot. The town had achieved such popularity that, despite all the new buildings, it was not possible to house all who wished to visit it.

There was plenty for the people to do after they had partaken of the waters. Each morning there was a large public breakfast at the 'Wells' and this was followed by music. Races were held daily at noon and in the afternoon there were cudgelling, wrestling and boxing matches, as well as foot races. The evenings were spent at private parties and assemblies or in card play.

Horse racing continued to be a popular sport. The famous race known as the Derby was first run in 1780 and the equally famous one known as the Oaks in the following year. (The first of these races was named after Lord Derby, and the second after one of his residences, the Oaks, which was situated near Epsom.)

Many people unable to visit the town were anxious to drink the famous water and arrangements were made to obtain crystals of magnesium sulphate from the water to send to them. These crystals fetched a very high price, as much as five shillings having to be paid for a tablespoonful.

In time, however, the popularity of the town waned. One reason was that crystals of the salt were being manufactured from material which had never been near Epsom. The substance, however, was used as a medicine for many years; and even today large quantities of Epsom Salt are still being sold.

9. *The Cave of the Dog*

THE CAVE of the dog, or, as the Italians call it, the Grotta del Cane, is situated on the shores of Lake Agnano, near the city of Naples. The lake is about two miles in circumference and occupies the crater of an extinct volcano. This cave has a very peculiar feature which is the reason for its name as well as for its popularity as one of 'the places to be visited'. Indeed, it has been visited for centuries, and the descriptions which follow were written in the eighteenth century. At that time the words 'fume' and 'vapour' were used for what we now call 'gas'.

From the floor of the cave, one description reads,

rises a thin, subtle, warm fume, visible enough to a discerning eye which does not spring up in parcels here and there but is one continuous stream covering the whole surface of the bottom of the cave and has this remarkable difference from common vapours that it does not, like smoke, disperse itself into the air but quickly, after its rise, falls back again and returns to the earth to a height of about ten inches. There is, therefore, no inconvenience by standing in it provided that the animal keeps its head above this level.[1]

But many unfortunate animals were not allowed to do so. Thus, we are told:

The person who conducted us to the Baths of St Germain is also the Guardian of this cavern. This man observing a dog in one of our carriages, was going to lay a hold on him, in order to try the usual experiment on that animal; but finding that I did not approve of this, he ran and fetched one of his own. Being returned with the creature, he entered the Grotta stooping a little, kneel'd down, and then seated himself upon his heels. He then took the dog by the four legs, and held it downwards, and this he did for some time. Immediately the dog howled, and fell into convulsions; then rolling his eyes and lolling out his tongue, all his nerves were contracted, after which he fainted away. In this condition, and struggling with death, his master threw him into the Lake of Agnano, but twenty paces from thence, when he immediately recovered himself, came out of the water, and ran away as fast as his legs could carry him; probably from fear of a second experiment. I asked the owner of the dog, whether he imagined that the water of the lake, singly, prevented such dogs as were taken out of the Grotta, from dying outright; to which he answered roundly, that it was owing to that water only, and that all Europe was of this opinion.

9. *The Cave of the Dog*

Whilst the guides' dogs knew what happened to them in the cave, and therefore had to be carried into it, a dog belonging to a visitor would willingly trot in behind its master. Inside the cave the men and women could walk about in safety but the dog, to the astonishment of the visitors, would almost at once fall to the ground at their feet.

Demonstrations were sometimes given on other creatures besides dogs. The well-known British author, Addison, describes what happened when a viper was taken into the cave:[2]

> A viper bore it nine minutes the first time we put him in, and ten minutes the second time. When we brought it out after the first trial, it took such a long vast quantity of air into its lungs, that it swelled almost twice as big as before and it is perhaps on this stock of air that it lived a minute longer the second time.

When Charles VIII, King of France, captured Naples in 1494 the cave fell into his hands, and one day he decided to make an experiment on an ass. The animal was led into the cave and forced to lie down on the floor. In a short time the donkey showed the same symptoms as the dogs and quickly died.[3]

This grotto has also been the place of death of a few men. Thus it is said that the Emperor Tiberius sent two slaves into the cave to their death. They were taken inside and chained to the floor and perished almost immediately. Later, it is reported, Peter of Toledo, Viceroy of Naples, shut up in the cave two condemned men, both of whom died.

In the 16th century a captured Turk was forced to lie on the floor, on the instructions of a Viceroy of Naples, in an experiment to see how long the slave would live. The men put his head below the 'vapour' and held him there too long, and 'the hapless Turk never recovered though plunged several times into the lake adjoining the grotto'.

It has been estimated that a dog dies in the grotto in three minutes, a cat in four, a rabbit in seventy-five seconds. 'A man', so it is said, 'could not live more than ten minutes if he were to be down upon this fatal ground.'

The eighteenth century illustration shows some of the incidents mentioned in this chapter. The man in the foreground can be seen

The Cave of Dogs and Lake Agnano

'reviving' a dog by throwing it into the lake whilst another dog is either running away after being 'revived' or else trying to escape the ordeal of having to go into the cave. The donkey is being forced to enter the cave.

* * * * *

These stories illustrate the fact that man and other animals cannot live in an atmosphere rich in carbon dioxide. An analysis of the air near the ground in the cave showed that it was composed of about seventy per cent carbon dioxide, six to seven per cent oxygen and about twenty-three per cent nitrogen. Ordinary air contains less than one per cent of carbon dioxide. Experiments with animals have shown that death may follow exposure to air containing more than twenty-five per cent of this gas but that air containing less than about ten per cent is not poisonous, unless it is breathed for a long time. Carbon dioxide is about one and half times as heavy as air and so stays near the floor of the cave. It is formed during chemical changes which are continuously taking place in the interior of the earth. Large quantities of it reach the earth's surface through volcanoes. The region of the Grotto is a volcanic one and there the carbon dioxide issues from the interior in very large amounts indeed, forcing its way into the cave through cracks in the rock floor.

10. The Republic has no Need of Scientists

ANTOINE LAURENT LAVOISIER, the son of a wealthy Frenchman, was born in the year 1743. As a youth he showed outstanding ability in his studies and was particularly attracted to the study of science, which at that time was rapidly gaining new students from every walk of life. His wealth enabled him to purchase the necessary materials and he quickly rose to be one of the most brilliant scientists of the time. At the youthful age of twenty-five, after he had made a geological survey of France in 1767, he was elected to the exclusive Royal Academy of Science of France.

The work which he did later more than justified this distinction. He showed the falsity of the theories then held about burning (or combustion) and helped to establish the use of an accurate balance as a necessary aid to science and all scientific investigations.[1] But since this story is concerned mainly with the relation between Lavoisier and the rulers of his land, special mention will be made only of his work for the government.

In 1775 he was appointed to take charge of the government's gunpowder factory and whilst a Commissioner for gunpowder invented means of increasing the explosive force of the powder. He also gave great help to the nation in the establishment of the metric system of measurement and in the application of science to agriculture. When the Revolution broke out his help was at first sought by the revolutionary leaders, who consulted him about manufacturing *assignats* (or paper money) of a kind which could not easily be forged.

Lavoisier in his laboratory

Before the Revolution the collection in France of some of the customs and taxes, such as those on tobacco, salt and a few alcoholic drinks, was 'farmed out' to a company of rich financiers called the 'general farmers'.[2] They paid an annual sum to the state but shared amongst themselves all the money they collected. In 1768 Lavoisier became one of these tax-farmers, and his ability soon won him a prominent place in the management of their affairs. He quickly amassed great wealth.

Tax collectors have never been popular in any age or country, but the tax-farmers in France were particularly disliked. Their main aim was to reap huge profits; they therefore enforced the regulations with the utmost rigour, and had tax evaders and smugglers, especially smugglers of salt, on which a high tax had to be paid, very severely punished. There were many scandals about their administration, especially when it became known that the tax-farmers made illegal payments to prominent and influential people; and that the King and his mistresses received large sums annually from them.

Few Frenchmen were surprised, therefore, when in 1791, two years after the Revolution, the National Assembly decreed the abolition of the tax-farm, and gave the tax-farmers two years to wind up their affairs. But the tax-farmers set about this job very slowly and had not reached a settlement by the end of the given time. Because of this unnecessary delay and for other reasons they suffered a great increase in unpopularity and in November 1793 a deputy demanded the arrest of 'these bloodsuckers'. The Convention followed its usual custom and ordered the arrest of all the tax-farmers, including Lavoisier.

The prisoners were kept waiting for trial until May 1794, when they were brought before the Revolutionary Tribunal. After the customary individual interrogations, their trial began. The presiding judge was a man named Coffinhall who was accustomed to make sarcastic and witty remarks at the expense of the victims appearing before him. The farmers were jointly and severally charged with practising all kinds of extortions and misappropriations at the expense of the people of France. They were also charged with taking excessive interest on sureties; with retaining funds which should have been paid to the Treasury; and with adding water and other ingredients to tobacco, thus making it harmful to the health of the citizens. The last-named charge was a trumped-up one, for the 'prosecutors' knew that some water had to be added to the tobacco-leaf during the manufacturing process; they could give no proof that the necessary amount had been exceeded or that harmful ingredients had been added.

Lavoisier and most of the tax-farmers were condemned to death, and, in accordance with the practice of those days, the sentence was carried out within a few hours of its pronouncement. During the trial an attempt was made to point out the great scientific service which Lavoisier had rendered France, but to no avail. A request was also made, so it is said, either by Lavoisier or by some person acting on his behalf, that the sentence might be deferred for a fortnight to enable him to finish some important experiments. It was at this point of the trial, it is alleged, that Coffinhall made his now notorious retort: 'The Republic has no need of scientists, justice must take its course.'[3]

Lavoisier's death came as a great shock to the intellectual world, Carlyle wrote:

> The Spring sends its green leaves and bright weather, bright May, brighter than ever: Death pauses not. Lavoisier, famed chemist shall die not live: Chemist Lavoisier was Farmer-General Lavoisier too, and now 'All the Farmers-General are arrested', all; and shall give an account of their moneys and incomings; and die for 'putting water in the tobacco' they sold. Lavoisier begged a fortnight more of life to finish some experiments; but the Republic does not need such; the axe must do its work.[4]

Lavoisier was executed only a few months before the end of the Reign of Terror, when Robespierre and many other leading revolutionaries, including Coffinhall, were also hurried to the guillotine. Gradually people in France began to feel safe in saying what they thought, and soon many of her scientists openly regretted Lavoisier's execution. It was then that Lagrange, the famous French scientist, spoke the now well-known words, 'It took but a moment to strike off his head; a hundred years may not suffice to produce another like it'.

On 12 August 1796, at the Lycée des Arts, a memorial service in honour of Lavoisier was held and the annual calendar of the Lycée has preserved a detailed account of a ceremony, the theatrical setting of which was in accord with contemporary taste.[5]

The entrance to the Lycée seemed to lead into a vast underground chamber and over it was inscribed 'To the immortal Lavoisier'. In the first rooms were replicas of the tombs of Voltaire and Rousseau, covered with wreaths, greenery and flowers. Facing the staircase was a pyramid twenty-five feet high, flanked by newly-cut poplars; its base was in the form of a funeral arch . . . of white marble and . . . inscribed: 'Respect for the Dead.'

The great hall had accommodation for three thousand and was draped in black, spangled with ermine, hung with garlands, and lit by twenty funeral lamps. On every column hung a shield bearing the title of one of Lavoisier's discoveries. At the back of the hall, where on either side stood replicas of the tombs of Desault and Vicq-d'asyr, hung a great curtain in the form of a ducal robe.

There was a vast audience, the men in black, the women in white wearing coronets of roses. The programme of the ceremony included a speech on 'The Respect due to the Dead', an eulogy of Lavoisier by the well-known scientist Fourcroy, with stanzas on the immortality of the soul. Finally came a cantata specially written for the service. For this singing the curtain at the end of the room was drawn and the principal singers, with a choir of one hundred people, appeared grouped around the 'tomb' of Lavoisier which was crowned by the statute of Liberty. And as the choir finished with the lines,

To render sacred his genius for ever
Let a monument be raised in his honour,

a pyramid appeared. On it was a bust of Lavoisier, the head encircled with the immortal crown of laurels which has traditionally always been conferred upon genius.

This memorial service must have been one of the most impressive services ever held in honour of a scientist.

11. Colours by Chance

A LATER chapter in this book deals with colour, but only with those colours which were sufficiently highly esteemed, centuries ago, to be reserved for the sole use of royalty and other important people. Today, there are numerous colours each of a brilliance and splendour not merely equal to but surpassing those ancient royal colours; and, what is more, they are produced at a cost well within the range of most families. A few of these dyes were discovered accidentally, as the following two stories relate.

One day in 1710 a colour manufacturer named Diesbach was doing experiments in his laboratory using alum, a solution of iron salt, and cochineal (a red dye-stuff). To do one of these experiments he required a solution of a substance known as an alkali and borrowed a bottle of it from another chemist who was working in the same laboratory.[1] He poured the alkali into the vessel containing the other substance and expected to get a red dye-stuff precipitated, as he had done in a previous experiment. To his utter amazement he got a blue one.

He discussed this strange colour-change with the other chemist, who recalled that the bottle did not contain fresh alkali but some which he had used in another experiment. In this other experiment the alkali had been mixed with some animal matter, probably bullock's blood,* and then it had been poured back into the bottle. Hence it was more than likely that the alkali was contaminated.

The second chemist, of course, should not have given a contaminated substance to the first one to use without telling him that the alkali might not be pure. Fortunately, in this case the careless act provided the chemists with a clue which they both realised was well worth investigating. So they deliberately repeated the former accidental occurrence. They added bullock's blood to a solution of pure alkali which they then poured on to another mixture containing cochineal, alum and the salt of iron. Again, a deep blue substance was formed. This they soon discovered to be a very good dyestuff. In honour of the country in which it was discovered, the substance became known as Prussian Blue.

For a few years its method of manufacture was kept secret and the substance was sold at a high price. But in 1724 a full description of the process was published by another chemist

* Diesbach's method of making the pigment called Prussian Blue was kept secret but it is thought that the 'animal matter' mentioned here was bullock's blood. This substance was used to obtain the dye a few years later, in 1724, by Woodward, the English chemist. He heated a mixture of dried bullock's blood and potassium carbonate strongly and obtained potassium cyanide. He next added green vitriol and alum to a solution of the cyanide, and when he treated the product with hydrochloric acid obtained Prussian Blue.

and before long the dye was being made in large quantities in many countries.

* * * * *

Before the next story begins it should be explained that until well into the nineteenth century chemical substances were divided into two groups, organic substances and inorganic substances. Organic substances were substances connected with life, such as sugar and starch, and it was believed that they could be formed only by Nature herself in the living body of a plant or animal. Inorganic substances, on the other hand, were those found in the ground, in the air or in the water, such as common salt, all the metals, oxygen, clay, limestone and the like, and they included all those substances which the chemist could then make in the laboratory.

In 1828 a German chemist discovered a laboratory method of making urea, a substance which before then had been made only in the body of an animal. This was a most important discovery and chemists gradually realised that they could make in the laboratory many other substances which up to then had been obtained from plants and animals only.

In a few years' time many other organic substances had actually been made in the laboratory; the German chemists, in particular, paying much attention to this side of chemistry. One of the most distinguished of these chemists was August Wilhelm Hofmann, whose special study was coal-tar. As a student, he had proved that a substance which had been extracted from tar was the same as one obtained from the indigo plant (which then yielded all the vast quantities of the blue indigo dyes which industry required). This substance was given the name of aniline from the two Arabic words *an-nil* meaning a blue substance.

In the year 1845 he was appointed professor at the Royal College of Chemistry in London. There he continued his experiments on making natural substances in the laboratory. Four years later he was speculating about making a natural substance called quinine but never actually began any practical work on it.

Quinine was then a most important drug and indeed it was prescribed frequently by most doctors as recently as the first half of the twentieth century. In 1849 it was obtained from the cinchona tree.

About three years later Hofmann appointed William Henry Perkin, a youth of fifteen, as assistant in his laboratory. Perkin had become extremely interested in chemistry at school and his parents encouraged him in his studies by allowing him to fit up a laboratory at home. There he spent most of his evenings after returning from the Royal College. Perkin's own laboratory was very simply equipped, as this description indicates:

> My own first private laboratory was half of a small but long shaped room with a few shelves for bottles and a table. In the fireplace a furnace was also built. No water was laid on or gas. I used to work with old spirit lamps, and in a shed I did some combustions with charcoal. It was in this laboratory I worked in the evening and vacation time.[2]

Not surprisingly Perkin was most interested in his master's research into the making of natural substances in the laboratory, and decided to try to make quinine during the Easter holiday of 1856. It has been suggested by some writers that Hofmann encouraged him to make the attempt. But Perkin, many years later, said that he was influenced by reading about the importance of quinine as a drug.[3] It may well have been, however, that Perkin had also learned about the speculations which Hofmann had made, seven years before, about the possibility of making quinine in the laboratory.

Perkin gave careful thought to his experiment before beginning the practical work. It seemed to him that a substance which Hofmann had just obtained from coal-tar was fairly similar in composition to quinine and so he decided to start with it. He considered the various other substances which would be required to convert it into quinine and then set to work. But his first experiments were unsuccessful.

Perkin then decided to use a different substance and chose the one which Hofmann had obtained from coal-tar in his student days.[4] So he put some of this substance, aniline, into a tube and treated it with a few other carefully chosen substances as before.

11. Colours by Chance

Young Perkin in his laboratory

This time he got a black precipitate at the bottom of the tube. On examination he found that much of the precipitate dissolved in alcohol but, instead of getting as he had hoped a colourless solution of quinine, he obtained a beautiful purple liquid.*

Soon afterwards he found that this liquid had the properties of a dye; and later that this new dye did not readily fade when

* Perkin wrote: 'I was led to the idea that quinine might be formed from toluidine by first adding to its composition C_3H_4, by substituting allyl for hydrogen, thus forming allyl toluidine.'

He prepared the allyl toluidine by the action of allyl iodide on toluidine, thus:

$$\underset{\text{toluidine}}{C_7H_9N} \text{ minus H plus } C_3H_5 = \underset{\text{allyl toluidine}}{C_{10}H_{13}N}$$

This he converted into a salt which he treated with potassium dichromate, expecting this reaction

$$2\, C_{10}H_{13}N + 3 . O = \underset{\text{quinine}}{C_{20}H_{24}N_2O_2} + H_2O$$

All he got was a dirty-looking precipitate. He wrote: 'Unpromising as this result was, I was interested in this action, and thought it desirable to treat a more simple base in the same manner. Aniline was selected and its sulphate was treated with potassium dichromate; in this instance a black precipitate was formed.'

This black precipitate contained about 5% of the substance which he called mauve and is now known as mauveine. It is a mixture of a few of the dyes of the phenyl phenazonium class.

exposed to bright sunlight. He showed it to a friend 'who, from his artistic taste, had a great interest in colouring matters'; he thought it might be a valuable dye-stuff.

Perkin, through another friend, got an introduction to Messrs Pullars of Perth, one of the most famous firms of dyers in the British Isles. He sent them specimens of dyed silk and received the following reply:

If your discovery does not make the goods too expensive it is decidedly one of the most valuable that has come out for a very long time. The colour is one which has been wanted in all classes of goods, and could not be obtained fast on silks, and only at great expense on cotton yarns. I enclose you pattern of the best *lilac* we have on cotton – it is dyed only by one house in the United Kingdom, but even this is not quite fast, and does not stand the tests that yours does, and fades by exposure to air.[5]

The delight and pleasure which such a complimentary letter from an eminent firm like Messrs Pullars gave to a youth of school age can well be imagined. Perkin persuaded his father and brother to help him and in a few months had started to build a factory to make the first dye ever obtained from coal-tar.

Perkin in old age in front of his first dye factory

The new dye was an immediate success, for, of all the colours in the spectrum, Perkin could not have chosen deliberately to discover one which would have had a greater appeal to the popular imagination of that time than purple. It is the colour which has been esteemed above all others for centuries – from the time when the famous Tyrian purple was reserved for royalty and the highest people of the realm – that is, even before the days of Emperor Nero. And by a curious coincidence, at the very time Perkin was discovering his purple dye the Empress Eugénie in France had started a fashion, which was quickly and enthusiastically followed by the ladies of England, of wearing purple dresses, which colour the French called mauve.

Perkin also named his new dye 'mauve'. The word soon achieved the greatest height of popularity – mention in the Victorian music hall, and the following account is interesting whether or not this popularity derived from the colour of the French dresses, as some say, or from Perkin's new dye.

Those only who were alive at the time know how the dye and the fact that it was derived from coal took the popular imagination. It was the topic of conversation everywhere, so much so that in a particular pantomime of the period, one of the characters, complaining of the way in which everyone would talk to him of nothing but mauve, added 'Why even the policeman says to you *mauve on* there'.[6]

In the course of the next few years other new dyes were obtained which quickly replaced the naturally occurring dyes which had been used for centuries. These new artificial dyes could be manufactured more readily and more cheaply than the natural ones and gave a greater variety of colour. These circumstances led Professor Hofmann to make this forecast, only six years after Perkin's discovery:

England will, beyond question, at no distant day, become herself the greatest colour-producing country in the world; nay, by the strangest of revolutions, she may ere long send her coal-derived blues to indigo-growing India, her distilled crimson to cochineal-producing Mexico, and her fossil substitutes for quercitron and safflower to China, Japan, and the other countries whence these articles are now derived.[7]

(Quercitron is a bark and safflower a flower; in 1860 both were used in the production of dye-stuffs.)

Unfortunately the British manufacturing chemists did not grasp the opportunity as did the Germans. Hence by the beginning of the First World War the Germans had a much better chemical industry than England. As a result of the war, however, Great Britain and other countries had learned of the numerous advantages of a flourishing chemical industry to a nation at war. Today the manufacture of artificial dyes from coal-tar is one of the world's greatest industries.

Hofmann's forecast that coal-derived blue dyes would soon be sent to indigo-growing India was rapidly fulfilled. Before long artificial indigo was being manufactured on such a large scale and so cheaply that the import of natural indigo practically ceased, and thousands of Indians, who had previously worked on the indigo plantations, were thrown out of work.

Another artificial dye called alizarin was made from coal-tar soon after mauve and it also soon replaced a natural dye which had been used for ages. This natural red dye was obtained from the roots of the madder plant, the growing of which had been a rewarding occupation for the farmers of France and other European countries. Its replacement was a great calamity for them.

Truly the discovery of the eighteen-year-old youth in his home-made laboratory during his Easter holidays started a great revolution in the chemical industry and in agriculture. It was a revolution, however, which probably would not have been long delayed even if Perkin had not made his lucky discovery in 1856.

12. *The First Balloons*

STEPHEN AND JOSEPH Montgolfier were two brothers who owned a large paper factory in Annoyan, a town on the banks of the River Rhône. Both were interested in the study of flying.

It seemed to them that a large paper bag filled with vapour and made as 'light as a cloud' would float in the air just as a cloud does. On the 5 June 1783, a large crowd gathered to see the brothers carry out an experiment to test the idea. A spherical paper bag thirty-five feet in diameter was fastened to the top of a long pole and a pile of straw and wood was built beneath the bag's open mouth. The pile was then set on fire and the smoke rose into the bag, which soon swelled into a large ball. On being released it rose so rapidly into the air that in less than ten minutes it had reached a height of about six thousand feet. Soon, however, it began to descend, and finally landed in a vineyard without damaging any of the vines.[1]

Professor Charles, a distinguished French scientist, heard of this remarkable experiment and decided to repeat it, with one important change. He knew that Cavendish, the British scientist, believed that a newly discovered gas, now called hydrogen, was ten times lighter than air. (Actually it is fourteen and a half times lighter.) So Charles decided to use hydrogen instead of hot air and smoke. At that time hydrogen was made in the laboratory from iron and dilute sulphuric acid.

Charles announced his plans and sought public subscriptions towards the cost of the necessary materials. With the help of two brothers named Roberts, he made a large 'globe' of silk about thirteen feet in diameter and glazed it inside with gum in order to make it gas-tight. The hydrogen was made by using about one thousand pounds of iron and five hundred pounds of sulphuric acid. These were placed in a vessel specially made to hold them, pipes being led from the vessel into the mouth of the bag. The silk bag was called a *ballon* or *balloon*: both words mean 'a great ball'.

Of course, such an experiment attracted much notice and many people gathered to see it when the filling commenced on August 23rd. Indeed, the crowds grew so great that the balloon had to be moved to an open space, called the Champ de Mars, some two miles away. It was taken there secretly by night. One eye-witness gave this account of its removal and flight:

> No more wonderful scene could be imagined than the balloon being thus conveyed. Men accompanied it with lighted torches and a detachment of foot soldiers and horse guards marched alongside it. This march by night,

the shape and size of the silk 'globe' which was carried with so much precaution, the silence that reigned, the unreasonable hour, all tended to give a singularity and mystery truly imposing to all those who were unacquainted with the cause. The cab-drivers on the road were so astonished that they stopped their carriages and knelt humbly, hat in hand, whilst the procession was passing.

The balloon is moved by night

Next day, the crowd of people in the flying field, on foot and in carriages, was so immense that a large body of troops had to be called out to prevent disturbances. At five o'clock in the afternoon, a signal having been given by the firing of a cannon, the cords that held the 'globe' were cut and it rose, in less than two minutes, to a height of nearly three thousand feet. It there entered a cloud, but soon reappeared, ascending to a much greater height; and at last it was lost among other clouds in heavy rain. The idea that an object was travelling in space was so sublime and differed so greatly from ordinary sights that all the spectators were overwhelmed with enthusiasm. The satisfaction was so great that ladies in the latest fashions allowed themselves to be drenched with rain, to avoid losing sight of the 'globe' for an instant.[2]

Fastened to the balloon was a leather bag containing a paper on which was written the day and hour of the ascent, with a request that the bag, if found, should be returned to Professor Charles.

Charles believed that he had put enough hydrogen in the balloon to keep it afloat in the air for twenty to twenty-five days. But after a flight of three-quarters of an hour the balloon fell in a field near a village called Gonesse, fifteen miles from Paris. The silk had split for a length of about a foot. The balloon had been carried to a height of probably twenty thousand feet, and at that height the pressure of the air outside was much less than that of the hydrogen inside: the greater pressure of the hydrogen had therefore caused the fabric to split. The balloon had fallen to earth because of the leakage of gas through the split.

According to the following newspaper account the villagers were terrified by the sight of this strange thing falling from the skies:

It was seen to fall by two peasants who took it for a flying monster descending from heaven. This impression was soon strengthened because, owing to the speed with which it fell, it bounced several times on the ground before it came to rest. These movements convinced the peasants that it was a beast which was alive; therefore they dare not approach it but stood for some time hurling stones at it. After they had thoroughly torn the silk in this way, the balloon lay quite still, so the bolder man of the two crept up and noted with astonishment that the beast had opened its jaws. He thought it was much too dangerous to put his hand into the mouth, for fear of the monster's teeth, so he satisfied himself with peering cautiously into it. But the unpleasant smell of the hydrogen [it was impure] had not dispersed and made him draw his head back again. His comrade noticed this from afar and, thinking that the beast had bitten him, made off with all speed. But the other man cried out that no harm had befallen him and that the beast was dead but stank. So they both plucked up courage, tied the balloon to the tail of a mule which was grazing near by, and dragged it thus to the village. They stopped at the priest's door and his Reverence was requested to examine the magic beast. He discovered the leather bag and read on the paper inside it by whom the machine had been made, to what purpose and to whom the bag should be sent. So the two peasants were greatly rejoiced they could expect a reward for their fright and labour.[3]

Another account of the astonishment of the villagers reads:

On first sight it is supposed by many to have come from another world; others, more sensible, think it a monstrous bird. After it has alighted there is yet motion in it from the gas it still contains. A small crowd gains courage from numbers, and, for an hour, approaches by gradual steps, hoping meanwhile the monster will take flight. At length one bolder than the rest takes his gun, stalks carefully to within a foot, fires, witnesses the monster shrink, gives a shout of triumph, and the crowd rushes in with flails and pitchforks. One tears what he thinks to be the skin, and causes a poisonous stench; again all retire. Shame, no doubt, now urges them on,

The flying monster from heaven

and they tie the fallen balloon to a horse's tail and the animal gallops across country tearing it to shreds.[4]

* * * * *

The eminent American scientist and statesman, Benjamin Franklin, who saw this first ascent, wondered about using balloons in warfare, and wrote:

> 5,000 balloons, capable of carrying two men each, would not cost more than five ships of the line, and where is the Prince who can afford to cover his country with troops for its defence so that 10,000 men, descending from the clouds, might not do a great deal of mischief in many places before a force could be brought together to repel them?

He thus forecast the use of paratroops in war a century and a half before time.[5]

Some of the French wondered whether 'England, our rival, will seize upon it and perfect it before us and so usurp the rule of the air as she has usurped too long that of the sea'. But people in England wondered whether their natural bulwark against invasion, the English Channel, would now not prevent the enemy from reaching their shores.[6] Indeed, a famous cartoon was drawn in 1784 entitled, 'Montgolfier in the Clouds'; it shows the French inventor blowing soap bubbles, which were supposed to represent balloons, and saying:

O by Gar, dese be de grand invention. Dese will immortalise my King, my country and myself. We will declare war against our enemies; we will make dese English quake, by Gar. We will inspect their camp, we will intercept their fleet, we will set fire to their dockyards, and by Gar, we will take Gibraltar, in de air-balloon; and when we have conquer de English, den we conquer de other countries, and make them all colonies to de Grande Monarche.

* * * * *

There was an interesting sequel to these first balloon flights some years later – that is in 1792. The French had by then rebelled against their king, and on the 'memorable tenth of August' the Parisian mobs became completely out of hand. They attacked the Royal Palace, massacred the soldiers who were guarding the king, and finally took the king prisoner. He was lodged in prison; later, after a trial of sorts, he was condemned to death and guillotined.

Living in the royal palace on that 'memorable day' was Professor Charles, for the king had rewarded him for his scientific work with free lodgings there. The mob went about the palace killing almost everyone they found, and some of them came across Charles. They were about to kill him when he reminded them of their delight at seeing his balloons a few years earlier. Some of the mob recognised him and so his life was spared. He survived the Revolution, and lived until the year 1823.

* * * * *

Franklin, as had already been mentioned, prophesied the use of paratroops. Fighting in the air was foreshadowed shortly afterwards, when, in 1808, two men for the first time engaged in aerial combat. They realised, even in this first fight, that a weapon firing only one shot at a time was of little use. So each man had a blunderbuss which fired numerous shots at one pull of the trigger, just as the machine-gun of the two world wars was used to spray its numerous bullets over a wide region of the air in the hope that one at least of them would hit a vital spot.

The two men, M. de Grandpère and M. le Piquet, had had a quarrel over an actress and according to the ideas of those days such an affair could be settled only by a duel. This, they decided, should be fought in the air from two exactly similar balloons. When all the long preparations had been made, each duellist, with his 'second', stepped into the car of his own balloon. The two balloons were placed in such a position that when they ascended they would be about eighty yards apart. They were released in the presence of an immense crowd, and, helped by a moderate wind, had risen to a sufficient height when the signal to fire was given. M. le Piquet fired first but missed, and then M. de Grandpère fired at the other balloon and hit. The balloon collapsed at once and the car fell, gathering terrific speed as it neared the earth, so that it hit the ground with a frightening crash. M. le Piquet and his second were dashed to pieces whilst M. de Grandpère and his second sailed on until the balloon finally landed some twenty miles from Paris.

* * * * *

Scientists soon realised that the use of balloons would enable them to study the upper regions of the atmosphere.

In 1804, two French scientists, with much scientific equipment, sailed in a balloon in an attempt to discover whether the magnetic needle behaved in the same way at high altitudes as it did on the ground.[7] During this flight an incident took place which caused a young and ignorant shepherdess to believe that a miracle had happened!

The balloon had risen to a height of about 7,000 feet but the two scientists wished to go higher. So they threw overboard many of the things they carried. One of the articles thrown out was a crudely-made white wooden chair. It fell to earth and landed on a clump of bushes, narrowly missing a shepherdess in its fall. The story, as related by an eminent French scientist, tells of her astonishment at seeing such an object falling from the skies. She knew nothing of balloons, of course, and could think of only one explanation of such an unexpected occurrence. The angels must have sent the chair from heaven for her use! She removed it from the bushes but, noticing its poor workmanship, was puzzled. Surely, she thought, the angels made better furniture than this. The riddle was not solved until some days later when a news journal gave details of the flight and mentioned the articles which the two scientists had thrown overboard.

* * * * *

Charles' idea of filling a balloon with hydrogen was by no means a new one, for a professor in Edinburgh University, Joseph Black, had done this some years before.[8]

In 1776 Black learned of Cavendish's discovery of hydrogen and the thought occurred to him that a thin and very light bladder full of the new gas would be lighter than the air it displaced and would, therefore, rise when set free. The sight of a bladder rising of its own accord would be most amusing, he thought. So he invited a party of friends to supper and after the meal set free a bladder full of hydrogen. Hence these friends became some of the first people to see a balloon rise in air. They wondered how Black had performed the 'trick' of getting the bladder to the ceiling. One possible way is mentioned in this description of that evening's entertainment:

> Soon after the appearance of hydrogen, which he showed was at least ten times lighter than common air, Dr Black invited a party of his friends to supper, informing them that he had a curiosity to show them. Dr Hutton, Mr Clarke of Eldon, and Sir George Clarke of Pennicuik, were of the number.
>
> When the company had assembled, he took them into a room. He had the allantois [the bladder] of a calf filled with hydrogen gas, and, upon setting it at liberty, it immediately ascended and adhered to the ceiling. The phenomenon was easily accounted for; it was taken for granted that a small black

thread had been attached to the allantois, that this thread passed through the ceiling, and that some one in the apartment above, by pulling the thread, lifted it to the ceiling and kept it in position. This explanation was so probable that it was acceded to by the whole company; though, like many other plausible theories, it turned out wholly unfounded; for, when the allantois was brought down, no thread whatever was found attached to it. Dr Black explained the cause of the ascent to his admiring friends; but such was the carelessness of his own reputation, and of the information of the public, that he never gave the least account of this curious experiment even to his class; and more than twelve years elapsed before this obvious property of hydrogen gas was applied to the elevation of air-balloons by M. Charles in Paris.[9]

13. *Light from Smoke*

BEFORE THE discovery of coal gas many attempts had been made to provide adequate lighting in the most important thoroughfares of the large towns at night. Thus history records that early in the fourteenth century the Mayor of London ordered that lanterns should be hung out in the streets by householders every winter's evening between Hallowtide and Candlemass. Experience had by that century showed the necessity of lighting the streets at night, if for no other reason than that robbers, and other men with evil design, found the dark streets a profitable hunting ground.

In the year 1668 and again a few years later the inhabitants of London were reminded of this old order. It seems that the order was not widely obeyed (or perhaps the lamps did not give sufficient light) for, in 1716, the Council of the City ordered all householders whose houses fronted any street, lane or passage, to hang out every dark night one or more lighted lamps from six to eleven o'clock, under penalty of one shilling (at that time a considerable sum). Many householders obeyed the order by simply not drawing their curtains, so that the light from an oil lamp in the room shone on the street.

A few years later some of the main streets were lit with oil lamps erected by the city authorities. But the light of these lamps was very dim indeed compared with the present-day ones. Hence people who could afford to do so hired link boys to walk before them carrying lighted rushes.

An interesting discovery that gas could be used for lighting was made in the year 1739 when a clergyman named John Clayton saw that the water in a ditch by the roadside, two miles from Wigan in Lancashire, seemed to be on fire.[1] 'This water', he wrote, 'burned like brandy', and he added that the flame was so fierce that men could boil eggs over it! The reverend gentleman knew better than the rustics, who thought the ditch contained a special kind of water, and persuaded some of them to empty the ditch and dig down into the ground at its bottom. In time, to use his words, 'a spirit arose from the ground'. This 'spirit' was the gas given off from the seams of coal which occurred underground.

We are told that he filled a good many bladders with the gas. When he wished to amuse his friends, he took up one of these bladders, pricked a hole in it with a pin and compressed it gently near the flame of a candle. The gas which was forced out caught fire and continued to burn till 'all the spirit was forced out of the bladder'.

The story now takes us to the last years of the eighteenth century, when William Murdoch, a young Scotsman, was resident manager in Cornwall for the firm of Boulton & Watt (discussed in a later book). This firm was then making the stationary steam engine designed by Watt and was finding a good market for it in Cornwall, where it was used for pumping water out of the mines.

Murdoch was born in 1754, the son of an Ayrshire farmer and millwright; and it is said that he used to make coal-gas even as a boy. There happened to be a layer of the kind of poor coal known as shale just under the surface of the soil in one of his father's fields. Young William collected some of this and put it in his mother's teapot. He built a fire in a cave, put the pot on it and lit the 'smoke' as it came out of the spout. The 'smoke', which was, of course, coal-gas, burnt with a yellowish flame.

This story is perhaps no more likely to be true than the story of Watt and the tea kettle, or the stories about other kettles and inventors of steam engines (see chapter 42), but if it is, then Murdoch may have recalled his early experiment when his new job gave him a cottage in Redruth, a Cornish village in the centre of the mining region. It is a fact that about the year 1792 he decided to try to light up his room in the cottage by gas.[2] The story of how he did this was related many years later by Mr William Symons, then an old man.

'Murdoch', said Mr Symons, 'was very fond of children, and not infrequently took them into his workshop to show them what he was doing.' Hence it happened on one occasion that Symons, then a boy of seven or eight, 'was standing outside Murdoch's door with some other boys trying to catch sight of some special mystery inside'. (For they knew that Dr Boaze, the village doctor, and Murdoch had been busy all the afternoon.)

Murdoch came out, and asked one of the boys to run down to a shop near by for a thimble. On returning with the thimble, the boy pretended to have lost it, and whilst searching every pocket, he managed to slip inside the door of the workshop, and then produced the thimble.[3]

He there saw that Dr Boaze and Murdoch had a kettle filled with coal on the fire and that they had been burning the gas issuing from the spout. They took the thimble and made a few tiny holes in it; next they fastened a small tube to the spout and put the thimble on the end of the tube. They then lit the gas as it came through the holes; and 'it burnt brightly in a continuous jet'.

In this manner Murdoch learnt that coal gas gives a bright flame when it is under pressure. He soon invented a 'burner' which had small holes in it, and soon also he made the first gas works ever constructed. This historic gas works was erected in his own backyard. Instead of the kettle he used a specially made iron vessel, nowadays called a retort, and placed it over an open brick fireplace. He brought the gas into the house by a pipe led through a hole in the wooden window frame and then up to the ceiling of his room. When the fire beneath the pot had been burning for some time coal-gas was given off and was set alight at one of his new burners fixed to the open end of the pipe in his room.

13. Light from Smoke

Murdoch lights the gas

Murdoch tried to interest his employers in his new discovery but Watt was not enthusiastic about it. He advised him not to continue with his experiments on coal gas but to concentrate his efforts on work connected with the steam engine. Murdoch was greatly disappointed, but when in 1798 he was promoted to be manager of their works in Birmingham, he soon managed to persuade the firm to manufacture gas-making apparatus for sale. His great opportunity came in 1802, when peace was declared with France and great celebrations were held throughout the country. The firm of Boulton & Watt decided that they would celebrate this 'Peace of Amiens' by lighting up their works in Soho, Birmingham, with gas. The sight was a great novelty, as the following account shows. The writer says that he 'was one amongst those who had the gratification of witnessing the first public exhibition of gas illumination', and continues:

> The illumination of Soho works on this occasion was one of extraordinary splendour. The whole front of that extensive range of buildings was ornamented with a great variety of devices that admirably displayed many of the varied forms of which the gas-light was susceptible. This luminous spectacle was as novel as it was astonishing; and Birmingham poured forth its numerous population to gaze at, and to admire this wonderful display of the combined effects of science and art.[4]

*   *   *   *   *

In 1802 it was not possible for Murdoch 'to light the gas' with a match, for matches as we know them had not been invented. He used a tinder box with a steel and flint. The steel and flint were struck together and the sparks they produced were directed into a tinder box containing old rags of cotton or linen which had been partly burned so that they would take fire easily. When the rags had caught fire they were blown upon until they glowed. A long splinter of wood, with a 'head' of sulphur, was then put into the glowing mass; the sulphur took fire and set the wood alight.

In 1827 John Walker, a chemist of Stockton-on-Tees, made a match head of a mixture which he thought would act better than sulphur.[5] The composition of this mixture is not known for certain but it probably contained some sulphur, together with potassium chlorate and antimony sulphide. One day he was dipping a number of splinters of wood into this mixture and putting them on one side until the mixture had set. He picked one of them up but in doing so happened to drag it a little on the hearthstone. To his great surprise the match head burst into flame on its own, even though the hearthstone was cold.

He reasoned to himself that it had caught fire through the friction between the head and the stone. So he took another match and deliberately stroked its head on the hearth. The match caught fire. In this way, so we are told, the first efficient friction match was invented; it became known as a 'lucifer'. Walker manufactured these matches and sold them at a shilling a box of eighty-four matches. A small piece of sandpaper was included. The paper was folded in two and held tightly while the match head was drawn sharply down the fold until it caught fire.

* * * * *

Murdoch's method of obtaining light from coal-gas became popular, and although doubts and fears were expressed by some people, there were others who welcomed it. Numerous interesting stories are told of the gas lighting of that period.

One such story relates that in 1818 Murdoch was in Manchester and had accepted an invitation to visit a friend's house to dine. The night was moonless and the roads were in a very bad state. Murdoch therefore filled a bladder with coal-gas and in its neck put the stem of a churchwarden's clay pipe, having stopped up the open end of the stem. Whenever he wanted to light his way he placed the bladder under his arm, removed the stopper from the end of the stem, and pressed the bladder. The gas forced out was then set alight.[6]

Many people believed that the gas was on fire in the pipes all the way from the gas works to the burner. A cartoon of the day shows an Irishman saying 'Arrah, honey, if this man brings fire through the water [in a pipe], we shall soon have the Thames and Liffey burnt down and all the pretty herrings and whales burnt to cinders'.[7]

Another story is that the workmen engaged to instal gas-lighting in the Houses of Parliament wished to put the gas pipes some inches from the wall so that the hot pipes would not set the building on fire![8]

Even scientists and other eminent men could not bring themselves to realise that this new method had come to stay. Thus Wollaston, a most distinguished chemist, said that 'they might as well attempt to light London with a slice from the moon'. The great Sir Humphrey Davy inquired sarcastically whether they intended to use the dome of St Paul's for a gas holder; and Sir Walter Scott wrote to a friend that 'there is a madman proposing to light London with – what do you think?, why, with smoke'.[9]

Fifty years later, in 1873, the Shah of Persia visited London and was so greatly impressed by gas-lighting that he particularly asked to be taken round a gas-works.

Now the Persians of old worshipped many strange gods, including the God of Light, whose name was pronounced Merdock. At the gas works the Shah asked many questions and, in the course of the answers, mention was made that gas-lighting had been made possible by the work of a man named Murdoch. The Shah at once thought of Merdock, the God of Light of his ancestors of Assyria, Babylon and Persia. He thereupon declared that Merdock must have been reborn again, reappearing as the Scotsman, William Murdoch, and ordered that portraits of William should be sent to hang in the palaces of Teheran and Kasr Kadjar.[10]

* * * * *

Other persons besides Murdoch have been mentioned as the discoverer of the fact that coal gas can be used to give light, but it would appear that the credit of being the first to use it on a large scale, for example, for the lighting of a house, belongs rightly to him.

14. *The Parson, Soda Water and Mice*

JOSEPH PRIESTLEY, who is often referred to as the father of English chemistry, became interested in the study of science mainly by chance. He was the son of a Yorkshire cloth-worker and was educated for the Nonconformist ministry. His early education, therefore, was of the type now known as classical and he learned little science, if any, whilst at school. In the year 1767 he was appointed minister of a chapel in Mill Lane, Leeds, and lived in the neighbourhood of a brewery.

Beer is made in large vessels called vats from barley, hops and yeast, as described in a later book. The yeast causes the liquid to ferment, that is, to bubble as if it were boiling; the bubbling, however, is simply due to the fact that a gas, carbon dioxide, is being given off. As this gas is much heavier than air most of it settles in a layer on top of the liquid in the vat.

In what follows, when Priestley's own words are used[1] the modern names of chemical substances are substituted for those given by him.

Priestley puts the candle into the carbon dioxide

'It was in consequence of living for some time in the neighbourhood of a public brewery', wrote Priestley, 'a little after midsummer in 1767, that I was induced to make experiments on carbon dioxide gas.' He often looked in at the brewery and learned that the layer of gas on the surface of the fermenting liquid was generally from nine inches to a foot in depth, and that there was a constant and fresh supply of it. When Priestley put lighted pieces of wood or lighted candles into the thick layer of gas the flame was extinguished each time.

In those days a spa water often prescribed by physicians was obtained from the town of Pyrmont in Germany, where it bubbled out of a natural fountain. It sparkled like the briskest champagne, had a pleasant taste, a slight smell of sulphur, and contained iron in solution, as well as the carbon dioxide gas which caused it to sparkle. Pyrmont water, as it was called, was bottled and exported and fetched a good price in England.

One day, having this pleasant but costly Pyrmont water in mind, Priestley thought of a way of dissolving carbon dioxide in water. His method was a simple one.

He used two glass drinking vessels, one full of water, the other empty. He held the empty one as near the surface of the liquid as possible, and, holding the full glass a foot or so above the liquid, poured the water through the layer of gas so that it fell into the the empty glass. The water dissolved some of the gas on its way down.

He then reversed the position of the glasses so that he could pour the water which had just passed through the gas through it again, catching it, as before, in a glass below. This process he repeated a few times. In this way, as he afterwards wrote, in the space of two or three mintes he made a glass of exceedingly pleasant sparkling water which could hardly be distinguished from very good Pyrmont. He continued:

I continued to make my Pyrmont water in the manner mentioned, till I left that situation, which was about the end of the summer of 1768 . . . When I removed myself from that house I was under the necessity of making carbon dioxide for myself, and, one experiment leading to another, I by degrees contrived a convenient apparatus for the purpose but of the cheapest kind.

The substances he used were chalk and acid and he purified the gas by bubbling it into water. Thus, in 1772, he was once more making Pyrmont water in his new home. His recipe was to add to each pint of carbon dioxide water a few drops of a tincture containing a strong solution of iron, a little spirit of salt, some oil of tartar and a few drops of acid. The 'water' so obtained, he claimed, 'had the peculiar virtues of Pyrmont or any other mineral water and had the same brisk or acidulous taste'.

Priestley printed his recipe, claiming that although his drink was as good as Pyrmont water 'Pyrmont water will cost you five shillings but mine will not cost you one penny'.

Later, carbon dioxide was made by the action of acid on soda (sodium carbonate) and the 'plain' solution of carbon dioxide in water became known as soda water.

Soda water became a popular drink and for a time it was also used as a cure for scurvy, a very common and dreadful disease among sailors which caused many deaths. It was known that the eating of fresh vegetables prevented it, but the reason was misunderstood. It was thought that more carbon dioxide was produced during the digestion of fresh food than was produced from food which had been stored for a long time on board ship. Hence a possible cure seemed to be to give carbon dioxide to the sailors to make up for the shortage. This remedy was highly thought of by many of the leading physicians of the day, so the Board of Admiralty fitted two warships with Priestley's apparatus for making sparkling water. But the 'cure' was not a success.

A great success was gained in another way, however, for before the end of the century the taste of soda water was improved by the addition of fruit flavours. Before long the manufacture of plain aerated table water on a large scale in this country and of flavoured ones in the United States of America had begun.

Thus it happened that the aerated table water industry of this country and the soft drink industry of America owe their origin to the fact that a Nonconformist minister lived next door to a brewery in Leeds.

During his residence in Leeds, Priestley was presented with a large burning glass or lens twelve inches in diameter and with a focal length of twenty inches. When such a glass as this is held in the sun's rays it focuses them on one spot; and the heat at this spot on a summer's day can be very intense. Priestley decided to find the effect of focusing the sun's rays on most of the substances in his laboratory, one after the other. In doing so he made a remarkable discovery on the first day of August 1774.

Priestley and the burning glass

On that day he focused the sun's rays on red oxide of mercury and obtained a gas which had previously been unknown. He then examined this new gas (which later became known as oxygen) by putting a lighted candle into a vessel full of it. His description of what happened reads:

What surprised me more than I can well express, was that the candle burned in the gas with a remarkably vigorous flame; the strength and vivacity of the flame is striking and the heat produced is also remarkably great. I cannot recollect what it was I had in view in this experiment; but I know that I had no expectation of the real issue of it. If I had not happened for some other purpose to have had a lighted candle before me, I should probably never have made the trial; and the whole train of my future experiments relating to oxygen might have been prevented.

Priestley then experimented to see whether creatures could live in the gas. His favourite method was to use mice in his experiments, about which creatures he wrote:

For the purpose of these experiments it is most convenient to catch mice in small wire traps, out of which it is easy to take them, and holding them by the neck, to pass them through the water into the vessel which holds the gas. If I expect that the mouse will live a considerable time, I take care to put into the vessel something on which it may conveniently sit, out of reach of the water. If the gas be good, the mouse will soon be perfectly at ease, having suffered nothing by passing through the water. If the gas be supposed to be noxious, it will be proper to keep hold of their tails, that they be withdrawn as soon as they begin to show signs of uneasiness. Mice must be kept in a pretty exact temperature, for either much heat or much cold kills them presently. The place in which I have generally kept them was a shelf over the kitchen fire-place, where, as is usual in Yorkshire, the fire never goes out.

The mouse sits over water

On the 8th of this month (March, 1775) I procured a mouse, and put it in a glass vessel containing oxygen. Had it been common air, a full grown mouse, as this was, would have lived in it about a quarter of an hour. In oxygen, however, my mouse lived a full half hour; and though it was taken out seemingly dead, it appeared to have been only exceedingly chilled; for, upon being held to the fire, it soon recovered, and appeared not to have received any harm from the experiment.

For my further satisfaction I procured another mouse. It lived three quarters of an hour. But not having taken the precaution to set the vessel in a warm place, I suspect that the mouse died of cold. However, as it lived three times as long as it could probably have lived in the same quantity of common air, and I did not expect much accuracy from this kind of test, I did not think it necessary to make any more experiments with mice.

My reader will not wonder, that, after having ascertained the superior goodness of oxygen by mice living in it (and by other tests), I should have the curiosity to taste it myself. I have gratified that curiosity, by breathing it, drawing it through a glass syphon. The feeling of it to my lungs was not sensibly different from that of common air; but I fancied that my breast felt peculiarly light and easy for some time afterwards. Who can tell but that, in time, this 'pure air' may become a fashionable article in luxury. Hitherto, only two mice and myself have had the privilege of breathing it.

Later, Priestley showed that burning took place much more rapidly in oxygen than in common air and stated that:

though pure oxygen might be useful as a medicine, it might not be so proper for us in the usual healthy state of the body; for as a candle burns out much faster in it than in common air, so we might, as may be said, live out too fast, and the animal powers be too soon exhausted in this pure kind of air. A moralist at least, may say that the air which nature has provided for us is as good as we deserve.

Before describing these experiments, Priestley wrote the following interesting passage in which the word 'philosophical' would nowadays be replaced by 'scientific':

The contents of this section will furnish a very striking illustration fo the truth of a remark which I have more than once made on my philosophical writings, and which can hardly be too often repeated, as it tends greatly to encourage philosophical investigation; viz. that more is owing to what we call chance, that is, philosophically speaking, to the observation of events arising from unknown causes, than to any proper design or preconceived theory in this business.

In the year 1780 Priestley moved to Birmingham where he was still living when the French Revolution broke out. His sympathies were with the revolutionaries, whereas most Englishmen hated them. Priestley rapidly became most unpopular because of his views. In 1791 a mob burnt down his house, destroying all his possessions, including his scientific apparatus and papers, and he himself narrowly escaped with his life. From that time onwards he was shunned by most of his former friends and acquaintances. He therefore decided to emigrate to the new republic of North America where at that time there was more freedom of thought and speech than in England. There he remained until his death in 1804.

15. The Fair Lady turns Black

HARROGATE HAS been famed as a health resort since the days of the first Elizabeth, in whose reign the sulphur springs which occur there naturally were given wide publicity by a local doctor, who believed that the waters were of great medicinal value. By the end of the eighteenth century about two thousand people were visiting the baths each season (from July to September) – a big number in those days, when travelling long distances was by no means easy. They lodged in inns, many of which had grown from simple village inns into large and palatial establishments. The visitors hoped to be cured of 'all manner of maladies and diseases, both inward and outward'.[1]

The building which housed the spa water was designed to enable the people either to drink the waters or to have a bath in them. A doctor wrote this in 1794: 'With us, warm bathing is not so much used by way of a luxury as of a remedy, and at Harrogate almost never with the former intention.'[2]

The history of Harrogate's spa is, however, very short compared with that of cosmetics, for throughout the ages womenfolk have sought ways of adorning themselves with paint, powder, and various preparations concocted by the chemist. One such cosmetic used by many ladies about the beginning of the nineteenth century was first made in France about the year 1600 and was sold by many druggists under the name of 'blanc de fard'.* In England it was known either as magistery of bismuth because it contained bismuth or as 'pearl white' from the brilliant white sheen it gave to the skin. It was customary to use a hare's foot to put on the powder, as shown in the illustration.

A woman powders her face with a hare's foot

* This white cosmetic has many other names, such as *blanc d'Espagne*, *perlweiss*, etc. It is formed by dissolving bismuth carbonate in as little concentrated nitric acid as is needed and pouring the bismuth nitrate thus formed into a large volume of water. Its chemical name is bismuth oxynitrate or basic bismuth nitrate. Various formulae have been assigned to it, for its composition varies according to the method of preparation; some are $BiO_3 . NO_5 . HO$ or $Bi_2O_3 . 5N_2O_5 . 8\ H_2O$, or $Bi(OH)_2 . NO_3$

The white powder turns grey on long exposure; and when used as a cosmetic it has been known, occasionally, to cause spasmodic trembling of the face ending in paralysis.

15. The Fair Lady turns Black

An early nineteenth century writer who styled himself 'An Old Philosopher' told an amusing story about a lady who used this cosmetic whilst at Harrogate:

> It was the practice with those ladies who were particularly ambitious of possessing a white skin, to daub themselves with a preparation of the metal bismuth. It is represented on creditable authority, that a lady made beautifully white by this preparation, took a bath in the Harrogate waters, when her fair skin changed in an instant to the most jetty black. You may judge how much was her surprise at this unlooked for change; uttering a shriek she is reported to have swooned; and her attendants on viewing the extraordinary change, almost swooned too, but their fears in some measure subsided on observing that the blackness of the skin could be removed by soap and water. The lady soon recovered from her trance, and deriving some consolation from having the true state of things explained to her by her physician, although she was not very well pleased that people should have discovered the cause of her white skin.[3]

'If any ladies', the old philosopher continued, 'continue to use this preparation, I would advise them to take particular care that they do not sit too near a coal fire, for their features would assuredly grow dark and dusty.'

The writer gave the reasons for this startling change in colour, which, in terms of modern chemistry, runs as follows: The water from cold sulphur-springs smells of the gas hydrogen sulphide – a gas which every student of chemistry soon learns to recognise. This smell is due to the presence sometimes of the free gas, and sometimes of the salt known as sodium sulphide. The water from the Harrogate cold sulphur-springs contains about 0·21 per cent of sodium sulphide. But when the spring-water is exposed to the warmer air the sodium sulphide undergoes oxidation (into sodium thiosulphate), and hydrogen sulphide is formed.

A few compounds of bismuth are white, some are yellow but one or two are black. One of these black compounds is bismuth sulphide, which is easily made in the laboratory by bubbling hydrogen sulphide into a test tube containing a solution of a bismuth compound. Evidently the black compound can also be formed on the skin of a lady daubed with magistery of bismuth who takes a bath in the Harrogate spa waters, even though the solution of sulphides is very weak. Also, as the old philosopher remarked, it is formed when a woman who is painted with it sits near a coal fire which is giving off sulphur fumes.

16. The Colour-blind Chemist

ENGLISH SCIENCE owes much to sons of working men, and John Dalton (1766–1844), who became one of our most brilliant chemists, is a good example of such a man. His father was a hand-loom weaver in a Cumberland village and his mother kept a small shop to help support the family. John was sent to the village school and showed such progress that he became a teacher at the age of twelve, spending his spare time in studying the classics, mathematics and science. Later, he taught at a school in Kendal but did not remain there long, leaving to teach at New College, Manchester, in 1793. Then started his long connection with the Philosophical Society of Manchester.

Dalton's interests in science were very wide indeed, but his main contribution dealt with the atomic theory. His theory of the atom gave a reasonable explanation of many of the then known facts of chemistry, and his Laws of Chemical Combination laid a sure foundation to nineteenth-century chemistry. This work brought world-wide fame to Dalton. The King, Parliament, scientific societies and the universities showered honours upon him. But he remained quite a simple man all his life. This was largely because of his upbringing in a sincere Quaker family.

The Quakers are members of a Christian group known as the Society of Friends. They believe that everyone is a child of God and so all are members of one large family who should help each other and be at peace one with another. Quakers, therefore, refuse to fight; but when their country is at war they are among the first to volunteer for medical work at the front and for similar humanitarian duties. It is a religion without priests: any

Quaker can take an active part in the service, which is held in a simply furnished room known as the Meeting House.

The Society was started in the seventeenth century, when the pronoun 'thou' was in general use; Quakers persisted in using 'thou' as a familiar and brotherly form of address long after other British people had adopted the word 'you'.

In Dalton's time many men, as well as women, wore brightly coloured clothes, and rich people had elaborate and costly dresses. But every Quaker, to emphasise the belief that all were equal, wore the same kind of simple clothes. These were usually grey in colour; bright colours such as red and scarlet were avoided because they might serve as distinguishing marks to make their wearers stand out amongst their fellows and equals.

Bright colours, especially red, feature in many stories about Dalton because his eyesight was not like that of most people – he was colour-blind. Little was known about this defect of vision until he made a very thorough study of it and published his results in 1794. Most people, he wrote, can see that there are six different colours in the spectrum – red, orange, yellow, green, blue and purple; but he could not. Red, to him, seemed 'little more than a shade or defect of light' – grey or drab. Yellow, orange and green looked almost the same as red. But he said that he could distinguish blue and purple.[1]

Dalton was not aware that his eyesight was somewhat peculiar until one day in his boyhood when he was watching soldiers on parade. One of the boys with him remarked on the brilliant colour of the red coats of the soldiers but Dalton said that they looked the colour of grass. Thereupon all the other boys laughed at him derisively and this made him think that his eyes differed from theirs. But he was not finally convinced of this until he was twenty-six, when his attention was focused on some flowers. Much later in life he gave this account of what he saw:

> I accidently observed the colour of the flower of the geranium by candle-light in the autumn of 1792. The flower was pink, but it appeared to me almost an exact sky-blue by day; in the candlelight, however, it was astonishingly changed, not having then any blue in it but being what I called red. Not doubting but that the change of colour would be equal to all, I requested

some of my friends to observe the phenomenon, when I was surprised to find they all agreed that the colour was not materially different from what it was by daylight.[2]

In another statement about his colour-blindness Dalton wrote: 'Pink appears by daylight to be sky blue, a little faded; by candlelight it assumes an orange or yellowish appearance . . . Crimson appears a muddy blue by day and crimson woollen yarn is much the same as dark blue.'[3]

The defect in Dalton's eyesight led to an exchange of letters with one of his friends. Dalton had written: 'I boldly assert, with grave face, that pinks and roses are light blue by day and a reddish yellow by night, and that crimson is a bluish dark drab.' His friend made fun of this: 'I find by your accounts you must have very imperfect ideas of the charms which in a great measure constitute beauty in the female sex; I mean that rosy blush of the cheeks which you so admire for being light blue.' The letter then continued that if Dalton knew of a girl with such an exceptional complexion she would be 'a fitter object for show than for a wife'.[4]

Some of the stories connect his colour-blindness with his membership of the Society of Friends. One relates that he saw a pair of stockings in a shop window in Kendall marked 'silk, newest fashion'. He examined them and bought a pair for his mother, Dame Deborah, who he knew had never had silk stockings and always wore home-knitted ones. But when he gave them to her she exclaimed, 'Thou hast brought me a pair of grand hose, John, but what made thee fancy such a bright colour? I can never show myself at the Meeting in them'. John was greatly upset by his mother's remarks and replied that the stockings seemed to him 'bluish-drab and Quakerish in all verity'.

'Why,' said his mother, 'they're as red as a cherry, John.'

John did not believe her; nor did his brother, Jonathan, who was called in to give his opinion (for Jonathan was also colour-blind). There being two to one against her Dame Deborah called in one of her neighbours, who quickly settled the dispute, saying, 'Varra fine stuff, but uncommonly scarlety'.[5]

16. The Colour-blind Chemist

Dalton gives his mother the stockings

Many years later, when Dalton was preparing to visit Paris to meet many learned Frenchmen, he decided to dress himself up for the occasion. So he went to a tailor's shop in Manchester, where he was well-known, and ordered a suit from a cloth lying on the counter. The tailor was amazed, for he knew that Dalton was a Quaker, and the cloth was intended for making red hunting coats![6]

Dalton's scientific achievements while he was at Manchester brought him great fame. The University of Oxford conferred on him an honorary degree and on his retirement the government gave him a well-earned pension. Moreover, he was presented to the King at a levee, or royal reception. A well-known scientist made the necessary arrangements for this. But he knew that Dalton, being a Quaker, could not go in court dress, because he would then have to wear a sword.[7]

It was suggested, therefore, that Dalton should wear the robes of a Doctor of Laws of Oxford. These robes were scarlet, and Quakers did not wear dresses of such a bright colour, but in this case there was no difficulty because to Dalton the robes looked to be 'the colour of dirt' and he agreed to wear them. The scientist who arranged for Dalton's presentation wrote:

> The dress of a Doctor of Laws is rarely made use of, except at a University function; Doctor Dalton's costume attracted much attention and compelled me to gratify the curiosity of many of my friends by explaining who he was. The prevailing opinion was that he was the mayor of some corporate town come up to get knighted. I informed my inquirers that he was a much more eminent person than the mayor of any city and, having won for himself a name which would survive when orders of knighthood would be forgotten, he had no ambition to be knighted.

* * * * *

As a result of Dalton's study of the subject, the particular kind of colour-blindness from which he suffered became known as 'Daltonism'. He believed that the cause of it was that the liquid in the interior of the eye absorbed the red end of the spectrum. Thus some of the colours were stopped from reaching the retina and so the person suffering from this defect was not aware of their presence. In order to test this belief he expressed the wish that his eyes should be examined after his death.

A medical friend of his, Mr Ransome, undertook the post-mortem examination and took out one of the dead scientist's eyes. According to one account he then held the eye in front of his own whilst he looked first at a red powder and then at a blue one. Each powder looked its own colour.[8] Another account of the experiment states that Ransome extracted the liquid from the interior of the eye and put it on a watch glass; this he held first over a red powder and then over a green one. The two powders appeared in their natural colours. Ransome therefore concluded that the liquid in the interior of the eye did not cause the change in colour.[9]

17. A Chemist Dramas

FRIEDRICH AUGUST KEKULÉ, student of architecture, professor of chemistry and day dreamer, was born at Darmstadt,

Germany, in 1829. On leaving school he began the study of architecture at the university but soon changed to chemistry.[1] Later in life he attained world-wide fame for his work on the way in which atoms are put together to form molecules. His youthful interest in architecture may have inclined him to study the structure of molecules.

About the middle of the nineteenth century chemists allocated a number to each element to indicate its combining power or valency. For example, they gave hydrogen 1 unit of combining power, oxygen 2 units, nitrogen 3 and carbon 4 units. Kekulé was one of the foremost in this work and used little pictures like the ones given below to represent atoms (these have been called Kekulé's sausages).

A carbon atom, 4 units,

A nitrogen atom, 3 „ N

An oxygen atom, 2 „ O

A hydrogen atom, 1 „ H

He combined these pictures of atoms when representing a molecule; thus a molecule of carbon dioxide was shown as:

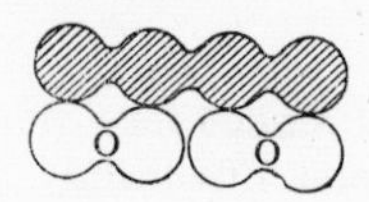

Most chemists, however, preferred a simpler method and used a short line for each link. Thus they wrote Kekulé's chain-link pictures for marsh gas, chloroform and carbon dioxide as shown below. The symbol of each atom has a number of lines attached to it equal to its number of units of combining power.

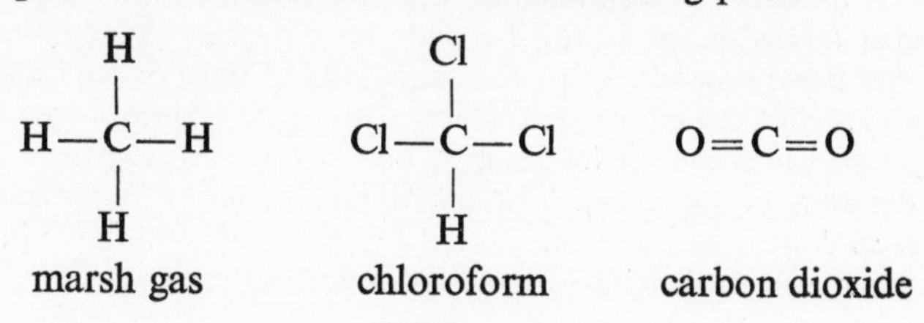

Kekulé had some difficulty in picturing the structure of molecules containing two atoms of carbon, such as that of ethane, which contains 2 atoms of carbon and 6 of hydrogen. For the structure must, therefore, show $2 \times 4 = 8$ links of carbon but only 6 of hydrogen.

He boldly tackled the problem by assuming that a molecule has, to use his words, 'the simplest and therefore the most probable structure'. Therefore the structure of a molecule of ethane may be pictured thus:

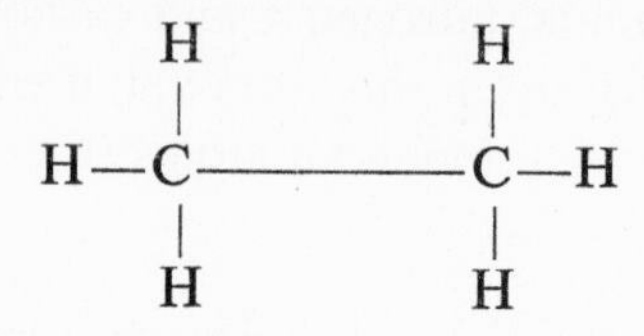

It will be seen that the joint link uniting the two carbon atoms is made up of one link from each carbon atom.

In 1854 Kekulé came to England as a visiting lecturer in chemistry and his ideas on the linking of atoms came to him whilst he was living in London. The following story, which he disclosed in 1890 in a speech before the German Chemical Society, describes what must be the most notable bus ride in the whole history of science:

During my stay in London I resided for a time in Clapham Road, in the neighbourhood of the Common. I frequently, however, spent my evenings with my friend, Hugo Miller at Islington, at the opposite end of the giant town. We talked of many things but oftenest of our beloved chemistry. One fine summer evening I was returning by the last omnibus, outside as usual, through the deserted streets of the Metropolis, which are at other times full of life. I fell into a reverie and lo, the atoms were gambolling before my eyes. Whenever, hitherto, these diminutive beings had appeared to me, they had always been in motion; but up to that time I had never been able to discern the nature of their motion. Now, however, I saw how frequently two smaller atoms united to form a pair; how a larger one embraced two smaller ones; how still larger ones kept hold of three or even four of the smaller; while the whole kept whirling in a giddy dance, I saw how the larger ones formed a chain, dragging the smaller ones after them but only at the ends of the chain. The cry of the conductor 'Clapham Road' awakened me from my dreaming; but I spent a part of the night in putting on paper at least sketches of these dream-forms.[2]

The smaller atoms of his dreams are those with only one link, the larger ones have two and the still larger ones three or four links. His two smaller atoms uniting to form a pair could have been, for example, an atom of hydrogen and one of chlorine, forming a molecule of hydrogen chloride as his 'dream pair', thus, H — Cl. In this molecule the single links of each of the two atoms combine, forming a joint one. Similarly his larger atom embracing two smaller ones could have been an oxygen atom embracing two hydrogen atoms, forming a molecule of water, thus H — O — H, whilst his still larger ones embracing three and four smaller ones could have resulted in the formation of a molecule of ammonia and marsh gas respectively, thus

```
   H              H
   |              |
H—N—H         H—C—H
                  |
                  H

ammonia       marsh gas
```

But Kekulé still had many problems to solve and in particular the compound known as benzene presented great difficulties. A molecule of this substance contains 6 carbon atoms, and thus a total of 24 carbon links, but only 6 hydrogen atoms with a total of 6 links. So the problem was to picture a structure in which as many as 24 carbon links were accounted for along with only 6 hydrogen links.

Once again the solution of the problem came to him in a dream one evening when he was sitting dozing before the fire at home in Ghent. This is the description which he gave of the dream.

> One evening I was sitting writing at my text book; but the work did not progress; my thoughts were elsewhere. I turned my chair to the fire and dozed. Again the atoms were gambolling before my eyes. This time the smaller groups kept modestly in the background. My mental eye, rendered more acute by repeated visions of the kind, could now distinguish larger structures, of manifold conformation; long rows, sometimes more closely fitted together; all twining and twisting in snake-like motion. But look! What was this? One of the snakes had seized hold of its own tail, and the

form whirled mockingly before my eyes. As if by a flash of lightning I awoke; and this time I spent the rest of the night in working out the consequences of the hypotheses.

The Chemist sits and dreams

'Let us learn to dream', continued Kekulé, 'then perhaps we shall find the truth, but let us beware of publishing our dreams before they have been put to the proof by the waking understanding.'[3]

In his dream his 'mental eye distinguished' long rows of atoms which were to become molecules of benzene. These are pictured below. In order to allocate 4 links to each carbon atom he had to include three double linkages between the carbon atoms as shown. It will be noticed, however, that one of the links of the first carbon atom and also one link of the last carbon atom in the row are shown, as yet, unattached.

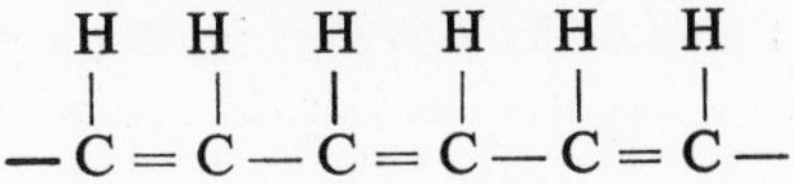

$$-\overset{\overset{\displaystyle H}{|}}{C}=\overset{\overset{\displaystyle H}{|}}{C}-\overset{\overset{\displaystyle H}{|}}{C}=\overset{\overset{\displaystyle H}{|}}{C}-\overset{\overset{\displaystyle H}{|}}{C}=\overset{\overset{\displaystyle H}{|}}{C}-$$

This and similar rows of atoms were the 'smoke-snakes' of his dreams which were twining and twisting. Then, in his dream, he saw that one of the snakes had seized its own tail, and this led him to join the unattached link of the first carbon atom to that of the last one. In this way he 'closed the chain' and obtained a ring of six carbon atoms all united together.

There are many substances which have characteristics in common just as members of a family have. Kekulé knew that each member of the benzene family, in nearly all its reactions, yields a chief product whose molecule contains at least six atoms of carbon. It seemed to him therefore that a molecule of every member of this family must contain a group of six united carbon atoms. This and other considerations led him to write the formula of benzene as shown below:

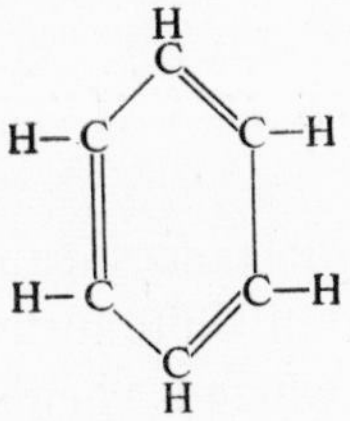

$$\begin{array}{ccccc} & & \overset{\displaystyle H}{C} & & \\ & / & & \backslash\!\backslash & \\ H-C & & & & C-H \\ \| & & & & | \\ H-C & & & & C-H \\ & \backslash & & /\!/ & \\ & & \underset{\displaystyle H}{C} & & \end{array}$$

This hexagon, which is known as the benzene ring or nucleus, indicates that the union of the six carbon atoms is so strong that it remains unbroken in all except the most drastic chemical reactions. The ring is frequently shown in a shorthand version, thus: ⬡

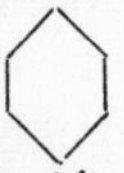

Kekulé's boldness in thus depicting the benzene ring was vindicated many years later when an X-ray examination and other modern tests clearly proved that the six atoms of the carbon in the benzene molecule are arranged in a hexagon.

The immediate result of Kekulé's theory was to enable chemists to give a reasonable explanation for much which they had previously found difficult to explain. But its final result has been of the utmost importance. For, by assuming the existence of the nucleus, chemists have been able to make or synthesise numerous other substances related to benzene, all of which contain this nucleus in their formulae.

Professor Japp, in concluding his Memorial Lecture on Kekulé, said this:

> The accuracy of Kekulé's predictions has done more for the deductive side of the science than that of almost any other investigator. His work stands pre-eminent as an example of the power of ideas. A formula, consisting of a few chemical symbols jotted down on paper and joined together by lines has supplied work and inspiration for scientific organic chemists during an entire generation, and affords guidance to the most complex industry the world has yet seen. Although much research remains to be done on the lines laid down by Kekulé, when this is fully accomplished no one will be entitled to more gratitude than August Kekulé.[4]

18. Tin is a Curious Metal

TIN, LIKE many other substances, can exist in different forms, for, as well as the common white lustrous silvery kind, there is also a rare grey powder which is chemically exactly the same as white tin. Indeed, when this grey powder is heated it changes into white tin; and conversely, white tin changes into a grey powder under certain conditions.

A striking example of this change from white to grey tin was noticed in the year 1851, when repairs were being done to a seventeenth-century organ in the castle church of Zeitz, a town in Silesia, Germany[1] where the winters are occasionally very severe. The organ pipes were made of an alloy containing 96·23 per cent tin and 3·77 per cent lead.

The workmen found that the pipe of the principal stop was covered with greyish, wart-like blemishes which looked like the marks or pustules left by smallpox on the face and hands. The damage to the pipe was extensive, one 4-foot length of it being covered with almost fifty of the wart-like swellings. These varied in size from a quarter of an inch or so in diameter up to about an inch. When the pipe was removed many of them crumbled into a grey powder.

Tin pox attacks the organ pipe

At first many scientists believed that the metal had crumbled into dust because of the vibration set up when the organ was being played. This belief, however, was not held for long because of another happening which occurred in a customs warehouse in St Petersburg.[2]

This incident was reported by an eminent Russian scientist in these words:

In February, 1868, I was informed by the head of a commercial firm here that a large number out of a stock of bars of tin in a custom's warehouse had disintegrated. I had a vague recollection that some years before a considerable stock of cast tin buttons made for the army and stored in a military depot, had, on inspection, been found no longer as such, but in their place was largely merely a shapeless decomposed mass, and that an inquiry into this quite inexplicable occurrence had been set on foot.

Since I am not aware whether this investigation had led to any result, I at once went to inspect on the spot the tin which had now been found decomposed. I found that, whereas a number of bars still appeared to be in their normal state, others had undergone a more or less radical change in their normal state.

From the beginning I had the strong impression that the cause of the change in the tin was to be sought in the exceptionally low temperature prevailing in St Petersburg in the winter of 1867–8.

Experiments made later confirmed the professor's impression, and he restored most of the grey powder to its former condition simply by heating it.

Today it is known that pure white tin, as 'the man in the street' knows it, is liable to change into grey powder at a temperature less than 13°C, and especially at temperatures around –40°C. Temperatures in most countries do not, of course, fall to the low level of –40°C, and the change, if it takes place at all, is a very slow one indeed.

One method of speeding up the change is to put a little of the grey powder on to a piece of white tin: then it may take place even at room temperature. Once it has started in one spot it quickly spreads until all the tin has 'sickened' and has become covered with pustules.

Tin workers have known about this disease for years; some call it the tin pox, others the tin plague or the tin pest.* The

* White tin is the ordinary lustrous metal and crystallises in a tetragonal form. Its density is 7·29 gm per cc., whilst that of the grey powder is 5·77 gm per cc. Hence, when white tin changes to grey, the volume increases by about twenty five per cent. When this large increase in volume takes place the 'excess' powder rises above the surface of the white tin and settles in the shape of a pustule.

The transition temperature, as found by an electrical method, is 13°C, but the rate at which it takes place is very slow, even at low temperatures. The transition is, however, greatly accelerated by the presence of even small quantities of grey tin.

The optimum temperature is about –40°C when the rate of change is at its maximum. But even at this temperature the rate is very slow; indeed, it may take years unless the white tin has been inoculated by the grey form.

The change is retarded by the presence of even small quantities of other metals, some of which prevent it altogether.

resemblance of tin pox to smallpox is most striking, for not only do the pustules closely resemble each other, but so also does the method of infection. For a person may get smallpox when pus is taken from a pustule on a person suffering from the disease and injected into his blood stream.

Tin warts or pustules are sometimes found on old medallions or coins containing tin and this fact has given the name 'museum sickness' to the disease. In all probability it has cost us many old medallions and coins.

* * * * *

It has been suggested that the tragic death of Captain Scott and his fellow explorers in the Antarctic could have been due to this change of ordinary tin to grey powder.[3]

Captain Scott, on his arrival in the Antarctic in 1911, followed the usual method of former explorers in setting up a line of depots stretching from the base by the sea to a place as near the pole as possible. A party of men was sent out before the wintry weather commenced with a store of food, fuel-oil, clothing and other necessities. They built depots at convenient distances apart in which they stored supplies. The largest of these was built about 150 miles from the ship and an abundant supply – one ton – of stores was placed in it. It was called One Ton Camp.[4]

When the time came to make the final dash to the pole, Captain Scott and four men set out from One Ton Camp, pulling sledges loaded with food and fuel-oil. At certain spots they set up small depots, leaving food and fuel in them in readiness for the return journey. They pushed on with all possible speed and eventually reached the pole. There a great disappointment awaited them. Flying in the breeze was a Norwegian flag which had been planted by Scott's rival, the Norwegian explorer Amundsen, who had travelled over a different route and had reached the pole a short time before.

The return journey was started in good weather, but before long the weather broke and conditions became appalling. The men had to face strong winds and blizzards and met with numerous crevasses in the ice which made the dragging of the sledges extremely difficult. Before long, one of the men contracted severe frostbite and died. The others journeyed on, calling at the small depots which they had set up on their journey to the pole. Here they found the food which they had left, but for some reason the quantity of oil was much less than

Scott finds the oil gone

it should have been.[5] However, they travelled on for another month, and then Captain Oates, who had been suffering painfully from frostbite for a long time, realised that he was incurable and was being a hindrance to his colleagues. So he walked out of the tent into a raging blizzard; he had decided to die so that his friends might have a chance of reaching safety.[6] But alas, his sacrifice was made in vain.

18. Tin is a Curious Metal

The remaining three explorers marched on. They reached a depot twenty miles from safety and replenished their stock of food and oil. Again the supply of oil was much less than it should have been. On they went for another nine miles: then another strong blizzard sprang up, so they decided to remain in camp under canvas for the day. But the blizzard lasted for days and it was impossible for any of the men to remain outside. It was clear to them that their fuel would become exhausted before their food. They knew that heat, especially a hot drink, is essential in the Antarctic. But when they had used all their fuel the weather was still as bad as ever; their fate was sealed. All three died; and they were actually only eleven miles from One Ton Camp and safety!

Captain Scott kept a diary and made this entry just before his death:

> I do not think human beings ever came through such month as we have come through and we should have got through in spite of the weather but for the sickening of Captain Oates and a shortage of fuel in our depots for which I cannot account and finally but for the storm which has fallen on us within eleven miles of the depot at which we had hoped to secure our final supplies.[7]

Many years after Scott's death an American chemist gave the following reason for the shortage of oil: 'The cans', he wrote, 'presumably were soldered with pure or too pure tin solder which did not stand the extreme Antarctic cold but turned into powder.'[8]

Such cans are normally made from tin-plate soldered at the seams, but the American had apparently not inquired about the purity of the actual solder used in this case. It is known, however, that the cold was extreme in the Antarctic that year, for Scott, in his diary, made special mention of the fact.

If any of the tin in the solder had turned into grey powder a tiny hole would have appeared when the powder fell out of the seam, as most of it would have done. If many holes had been formed an appreciable volume of oil would have escaped in the course of weeks.

There is no doubt that oil had passed out of the tins, for it is recorded that 'a search party found at One Ton Camp that some of the food, stocked in a canvas tank at the foot of the cairn was quite oily from the spontaneous leakage of the tin seven feet above it on top of the cairn'.[9] But whether the above explanation of the leakage can be accepted is another matter.

When this suggested cause of the leakage was first put forward all the information which is now available had not been collected. It was known that pure tin is much more liable to change than impure – indeed, the change is almost always confined to tin of the highest purity. In most of the ordinary kinds of solder, however, there is a substantial proportion of lead, the presence of which would go far towards preventing the change. Solder has been used for a long time in assembling the parts of many appliances used at low temperatures – refrigerators for example. But recorded cases of tin pest affecting these appliances are very rare indeed.

There is a very slight possibility that the tin in the solder used in making the cans was infected with grey powder before they left this country, and infected white tin, even when slightly impure, if kept at low temperatures for a long time is liable to turn into grey tin.

The assumption that tiny holes appeared in the tin was, however, contrary to an observation made by the leader of the team which set out in an attempt to rescue Scott and his party and stayed for a short time, whilst on their way towards the pole, at One Ton Camp. He wrote that, 'they found the stores left at this camp and noticed that one of the tins of paraffin on top of the cairn had leaked and spoilt some of the stores placed at the foot of the cairn. There was no hole of any kind in this tin'.[10]

This definite statement that there was no hole in the tin makes it difficult to accept the explanation suggested in this story as the cause of the leakage, unless it is assumed that the holes were so tiny as to escape notice.

Huxley, who edited Scott's diary, gave this totally different explanation:

As to the cause of shortage, the tins of oil at the depots had been exposed to extreme conditions of heat and cold. The oil was specially volatile, and in the warmth of the sun (for the tins were regularly set in an accessible place on top of the cairn), tended to become vapour and escape through the stoppers even without damage to the tins. This process was much accelerated by reason that the leather washers about the stoppers had perished.[11]

There is now, however, a very sound reason for rejecting the American's explanation.

In 1956 an expedition to the Antartic recovered some of the stores left by Scott forty-five years before and brought them back to England. A few oil cans were among the stores recovered. They were scientifically examined to test the stability of tin, and this statement was made afterwards:

A convincing example of the stability of tin when exposed to low temperatures is provided by the condition of the tinplate cans from the Scott Expedition to the Antarctic in 1911, which were recovered and examined in 1957 at the Tin Research Institute. No trace of grey tin either externally or internally was discovered when these cans were examined.[12]

19. Nobel – His Discoveries and his Prize Fund

IN THE year 1846 Professor Sobrero, an Italian chemist, discovered a new substance which was soon to replace gunpowder for many purposes. This was nitroglycerine, an oily liquid which is highly explosive in an unpredictable way. Thus, whilst normally an explosion occurs only when some of it is poured on to a hard surface and is then struck a blow,[1] at other times it will explode violently when the bottle containing it is given even the slightest jolt. Its discoverer knew of these properties and issued a warning that the oil would be dangerous if used for industrial purposes.

A moderately safe way of using it was, however, later discovered and the liquid became known as blasting oil, since it was used in quarrying and mining to blast the rock.

In 1860 Emmanuel Nobel, who had long been interested in explosives, decided to open a factory near Stockholm for the manufacture of nitroglycerine. His two sons helped him in the new venture. Unfortunately the business had a very tragic start: very soon after the factory was opened, the liquid exploded and shattered all the buildings, killing many workpeople, including one of Nobel's sons. But with the help of Alfred, the remaining son, Nobel started afresh, and soon they were producing nitroglycerine once again on a commercial scale.

The liquid was a very difficult one to transport because of its liability to explode when jolted or shaken.[2] Hence the tins in which it was despatched were placed close together in a wooden case and the spaces between them were firmly packed with sawdust. The nitroglycerine, however, contained an impurity which reacted with metal and so occasionally tiny holes were formed in the walls of the tin cans.[3] When this happened, the liquid which passed through them soon saturated the sawdust and then trickled out of the crate on to the roads and railway tracks as well as on to the clothes and boots of the men who handled the tins.

Later a substance called kieselguhr was used instead of sawdust. This is a white powdery substance which was formed from the remains of small sea creatures thousands of years ago when the land was under the sea. Large deposits of it occur in the neighbourhood of Nobel's factory which was situated near Hamburg. The substance is easily dug up, so there was a cheap and plentiful supply of a substance which could be used for packing purposes.

It is said that the substance had been in use for a short time only when a workman who was unpacking a crate noticed that although some of the liquid had leaked out of a tin none of it had trickled out of the crate; all had been soaked up, that is absorbed, by the kieselguhr. When Alfred Nobel heard of this, an idea for making better use of the kieselguhr than for packing purposes flashed through his mind. He tested the idea immediately and his experiments showed him that kieselguhr was a very porous kind of white earth which would absorb about three times it own weight of liquid nitroglycerine, becoming only slightly damp in the process. But he also found that this damp mass of nitroglycerine and kieselguhr had somewhat different properties from the liquid itself, the most important of all being that it was not sensitive to shock. Therefore it did not explode when shaken or jolted; indeed, it could even be burned in the open air without exploding. Nevertheless, it exploded with violence when a detonating fuse set it off. Nobel called it dynamite.

This oft-told story does not tally with Nobel's own account of the discovery of dynamite. He stated that he deliberately began experiments to find a substance which would absorb the liquid, and tried sawdust, charcoal, brickdust and other porous solids but met with little success. He then tried kieselguhr and found it was the most suitable substance for this purpose.

The new explosive, dynamite, was soon in great demand by engineers for the construction of mines, tunnels and roads and by quarrymen for blasting rocks and for many other purposes (including the blowing open of safes!), for which its pasty nature made it particularly suitable.

For many years, however, even though the nitroglycerine was sent out in its safe form as dynamite some transport difficulties were encountered. It is said that when railway companies refused to carry it 'iron-nerved salesmen' recruited from mines and quarries, packed it into trunks as "personal luggage", or in boxes labelled "Glass, Handle with Care". It was stored in hotel sample rooms as "China, Fragile", and hidden in cellars under beds'.

Nobel endeavoured to establish nitroglycerine factories in many countries but at first with little success.

> He went to Paris to secure financial backing for his invention. He told the French bankers that 'he had an oil that would blow up the globe'. But the bankers thought their interests lay in leaving the globe just about as it was. When he went to New York his luggage consisted of a few trunks of dynamite. He used to say that not a hotel would take him in and that the New Yorkers shunned him as though he had brought the pest in his pocket.[4]

Many such stories and anecdotes are told about him, some of which may have a little truth in them. In any event, Nobel did ultimately succeed in establishing factories in France as in most other countries, especially after his next discovery, which he made in 1875.

In that year, whilst experimenting with nitroglycerine, Nobel cut his finger and painted the cut with a substance known as collodion which was then commonly applied to cuts because it set in a few minutes into a kind of skin and thus kept dirt out of the open wound.

With this 'new skin' on his finger Nobel continued with the experiment, but happened to spill a little of the nitroglycerine. Some of it fell on the collodion. To his surprise he saw that the collodion had altered in appearance, and he was too good a scientist to let such an unexpected occurrence slip by without further investigation. Accordingly, he made a number of interesting experiments. In the course of these he found that when nitroglycerine is heated with finely divided collodion a gum-like substance is obtained. He then discovered that this transparent, jelly-like gum was a more powerful explosive than dynamite. Nobel began to manufacture this new substance and called it dynamite-gum, but later, to avoid confusion with dynamite itself, he gave it the name of blasting gelatin.[5]

This chance happening, which Nobel did not deny when he denied the story about dynamite, justifies the claim that 'blasting gelatin was born on a man's finger and not in a test tube'.

* * * * *

One of Nobel's friends, the Baroness von Suttner, had written a book entitled *Lay Down Your Arms*, which enjoyed great popularity amongst advocates of peace. She tried to get Nobel to help in her efforts to abolish war, and he was far from unsympathetic to her views. But he differed from her as to the best method of persuading all nations of the folly of war. Thus: 'I wish that I could produce a substance or a machine of such frightful efficiency for wholesale devastation that wars should thereby become altogether impossible',[6] and again: 'My factory may make an end of war sooner than your congress. One day when two army corps will be able to annihilate each other in one second, all civilized nations will recoil from war in horror and disband their forces.'

It is interesting that a little over half a century later, when such a weapon – the H-bomb – was invented many people, as Nobel had predicted, did recoil in horror from warfare on realising the terrible devastation which would result from a future major conflict. This feeling was expressed by no less a person than Mr Eisenhower, President of the United States of America and Supreme General of the Allied Forces in the Second World War, in these words which he broadcast on the last day of August 1959:

> When we are talking about peace we are talking about something that is imperative in our time. War has become so frightening in its capacity for destruction of the whole of civilisation that we – and I mean all people, as well as statesmen – have the responsibility of making sure that our actions in all things we try to do are all directed by this single purpose and directed with what intelligence we can marshal with the brains which the Good Lord gave us.

Nobel, however, did a great deal more for peace than making such a prediction, for he decided to leave most of his huge fortune of some millions of pounds to be used for the good of humanity. The money was to be used to endow prizes for those who had done the most to advance the idea of general peace and friendliness between nations, by their efforts towards the abolition or reduction of the strength of standing armies, by encouraging peace congresses or by otherwise rendering great service to mankind.

Nobel died in the year 1896 and the Nobel Prize Fund was established in 1901. Since that year prizes worth many thousand pounds each have been awarded annually from this fund to eminent persons of any nationality and of either sex. In accordance with the scheme, one prize is given to the person selected by the Norwegian Storthing (Parliament) as having done most during the year just ended to promote peace; and one is given, on the advice of the Swedish Learned Academies, for eminence in each of the subjects of medicine, chemistry, physics and literature.

20. *A Jewish Chemist Regains his Fatherland*

THE JEWISH people are descendants of the tribes of Israel who settled in Palestine many centuries ago. Their race has, therefore, a very long history. This story about a Jewish chemist begins in Jerusalem about six hundred years before the birth of Christ, when the city was strongly fortified and had within its walls the famous Temple of King Solomon. This temple was the centre of the Hebrew religion and its presence made Jerusalem a holy city.

In the year 586 B.C. Nebuchadnezzar, king of Babylon (a city in Mesopotamia) captured Jerusalem and destroyed it. Like some conquerors of modern times he sent many of the citizens into exile as slave labour, some of them being taken to Babylon. About fifty years later, when Cyrus, king of Persia, captured Babylon, he set free the deported peoples and their descendants, allowing any who so desired to go to Jerusalem. Many did so and settled in the ruined city under the leadership of Nehemiah.

Nehemiah immediately began to reconstruct the city. He rebuilt the walls and fortified them once more. He restored King Solomon's Temple, and gradually re-established the Jewish mode of life and worship. The city of Jerusalem once more became a holy city and remained so for the next five hundred years. Nehemiah, rightly, is still regarded as one of the greatest men of his race.

In the year 70 A.D., however, Jerusalem was again destroyed, this time by the Romans. Once more its people were made homeless, but this time their conquerors did not even trouble to transport them. The Jews were simply driven from home and left to find a dwelling place wherever they could. From this time onwards they were scattered throughout the then known world, a race of people without a fatherland. But no matter where groups of Jews settled they remained faithful to the religion of their forefathers. This, together with their common literature and language, has held them together as one race throughout the centuries.

During these years some of the Jews still looked towards Jerusalem as their holy city and preserved the hope that one day their people would once again dwell as a nation in Palestine, the land of their fathers. Towards the end of the nineteenth century a number of Jews who called themselves Zionists banded together with the aim of founding a national home for Jewish people in Palestine, although not all Jews were in favour of this policy.

The First World War of 1914–1918 found Jews fighting on the side of the country in which they lived, whichever it was. This story is about a Jewish chemist who helped the British; the story in chapter 21 tells of another Jewish chemist who helped the Germans.

At the outbreak of World War I, Chaim Weizmann, a Jew from White Russia, was employed as a lecturer at Manchester University, where he was experimenting on the production of artificial rubber. He was a keen Zionist, sparing no effort to further the cause of Jewry, but his great opportunity to help the Zionist movement came through his scientific work.

This work was on acetone. Acetone is a liquid which dissolves many substances and for that reason is much used in manufacturing processes. In war-time it is in great demand because of its role in the manufacture of cordite, which is the explosive used in rifle bullets and in many shells. Up to the year 1914 the usual method of manufacturing acetone was by heating wood in closed vessels (so that the air was excluded) and collecting the vapour which was given off. This vapour contained acetone. Large quantities of wood were required to make all the acetone needed.

There are, however, only a few large forests left in the British Isles, and so before the war almost all the wood required for the manufacture of this liquid had to be imported. But in war-time shipping space is very valuable, particularly when much of the cargo carried by ships is lost by enemy action. One way of alleviating the war-time shipping difficulty was to discover a method of making acetone from some substance other than wood; the substance had, of course, to be common in the British Isles.

Dr Weizmann had discovered such a method in 1910 whilst engaged in his experiments in making artificial rubber. In these experiments he had hoped to find bacteria (that is, germs) which would convert sugar into another substance which could then be converted into rubber. He did not succeed in doing this but he did discover, by chance, a germ which brought about the conversion of sugar into pure acetone.*

Acetone, however, could not be converted into artificial rubber and therefore Weizmann and his professor attached little importance to his chance discovery; indeed, the professor advised him to pour the stuff down the sink. Fortunately Weizmann did not forget the details of the experiment.[1]

In 1914, at the outbreak of war, the War Office sent a circular to scientists inviting them to report any discovery of military value. Weizmann reported his method of making acetone but no steps were taken about it for some time. By the end of the next two years, however, the acetone problem had become acute. Shipping could not easily be spared; and some people believed that the acetone obtained from wood was not pure enough for the manufacture of good cordite especially after the naval battle of the Falkland Islands, when, it was said, some of the shells fired from the British ships fell short of their target because the acetone used in the manufacture of the cordite was

* Chemists have known for a long time that the ethyl alcohol which is produced by the fermentation of sugar by yeast contains a little isoamyl alcohol as an impurity. Weizmann needed relatively large amounts of the last-named alcohol in his efforts to synthesise rubber and decided to see whether he could obtain it as the main product from sugar.

He decided to use a bacteriological method. By it he obtained a liquid which smelled like isoamyl alcohol. On analysis, however, he found that the liquid was a mixture of acetone and butyl alcohol and so he pursued the method no further at the time.

During the war he developed his manufacturing process from the result of his accidential discovery of acetone. The process consisted of the conversion of starch to sugar which was then treated with a bacterium. The bacterium came to be called closteridium acetobutylamine because the product consisted of about sixty parts of butyl alcohol, thirty parts of acetone and ten parts of ethyl alcohol.

not pure. Be this as it may, early in the war Dr Weizmann was asked to undertake the manufacture of pure acetone from sugar.

At that time the British Isles did not produce a large amount of raw sugar, most of it being obtained from the sugar-cane plantations of America or from sugar-beet from Europe.

The shells fall short

Nevertheless, Britain grew large quantities of wheat, barley and oats, as well as many potatoes. All these substances contain starch which can be converted fairly simply into a kind of sugar which is fit for manufacturing purposes.

Weizmann was seen by Mr Winston Churchill at the Admiralty in 1916. Some years later he confessed that he was terrified by almost the first words uttered at the interview. They were, 'Well, Dr Weizmann, we need thirty thousand tons of acetone. Can you make it?'[2] Up to that time Weizmann had only made about as much as would fill a tea cup! He knew, all too well, the difficulties of converting a method used in the laboratory into a large-scale manufacturing process.

Mr Lloyd George (then Chairman of the Munitions of War

Committee), who was largely responsible for enlisting Weizmann's help, also interviewed the chemist.[3] Dr Weizmann was cautious, and while he admitted that he could produce acetone in the laboratory he stressed that he would require some time before he could guarantee successful production on a manufacturing scale. 'How long can you give me' he asked. Lloyd George said, 'I cannot give you very long. It is pressing'. To this Weizmann replied, 'I will go at it night and day'.[4]

He was allowed the use of Nicholson's gin distillery in Bromley-by-Bow, and, after numerous difficulties, found a method of making about half a ton of acetone at a time from sugar which was obtained mainly from maize. The Admiralty then took over many other distilleries and also built a new one for him. Soon the factories were using half a million tons of maize a year. All this had to be imported from America but it took up less shipping space than wood.

At this stage of the war, however, the German U-boats were torpedoing so many of our ships that finally substances other than maize had to be used. Substances containing starch could ill be spared and to supplement the scarce supplies the children of Britain were asked to collect chestnuts, the starch of which could be made readily into sugar.[5] Before long, however, the government decided to set up acetone factories in Canada and America where there was much more starch available than in England, for there was ample maize and similar grain. Factories were also established in India, the starch there being obtained from rice. Before the end of the war the Allied factories were making acetone of a pure quality in sufficient quantities to meet all the needs of war-time.

By this time Weizmann's scientific work had brought him into close contact with some of the leading statesmen of this country. It is true that before the war he had met Mr Balfour, who later was to become the war-time Foreign Secretary.[6] But the acquaintanceship was very slight. In 1916 Weizmann and Balfour met again to discuss official business connected with the production of acetone. Towards the end of the interview Balfour, knowing of Weizmann's intense keenness in Zionism, began discussing the Jewish problem and concluded 'You know, Dr Weizmann, if the Allies win the war you may get your Jerusalem'.[7]

Some months later Lloyd George, then Minister of Munitions, sent for Weizmann to congratulate him on the success of his efforts in manufacturing acetone and said: 'You have rendered great service to the State and I should like to ask the Prime Minister to recommend you to His Majesty for some honour'. To this Weizmann replied: 'There is nothing I want for myself.' This took Lloyd George aback and he said: 'But is there nothing we can do as a recognition of your valuable assistance to the country?' Then Weizmann said boldly: 'Yes, I would like you to do something for my people'. He went on to explain that he was a Zionist and therefore passionately hoped that Palestine would after the war be restored to the Jews as a national home.[8]

Lloyd George was deeply impressed but at the time could do little to help the Jews. He did, however, discuss Weizmann's work and hopes with Balfour, who was very interested in scientific matters as well as political ones. Soon the Jewish chemist and the Foreign Secretary had become closely associated. Little progress was made, however, until Lloyd George became Prime Minister, when after long negotiations with leading Jews the famous Balfour Declaration was approved in November 1917. It read: 'His Majesty's Government view with favour the establishment in Palestine of a national home for Jewish people and will use their best endeavours to facilitate the achievement of this object.'

This declaration was issued with the knowledge of the other Allied Powers, all of whom shortly afterwards approved it.

Palestine was in the possession of the Turks before the First World War. In that war they fought against Britain and her Allies. Towards the end of 1917 General Allenby, who was in charge of the British Forces in the Middle East, launched a highly successful attack against them. His forces advanced so rapidly that the Turks in full retreat had no time to lay waste the land or sack Jerusalem. So it happened that only a week after the Declaration was agreed by the British statesmen Allenby made his triumphant entry into an unspoiled Jerusalem. The Jews from the allied nations could now visit their national home in safety. So a Jewish Commission, authorised by the British Government, and headed by Weizmann, was sent to Palestine to deal on the spot with various matters arising out of the Declaration.

Thus for the first time since the year 70 A.D. the conditions seemed favourable for the rebuilding of the Jewish national home in Palestine. By the year 1920 thousands of Jews had settled there. More and more of the land was brought under cultivation,

Weizmann is elected first President of the State of Israel

industries were started, schools set up and a university founded. The way was not smooth but after many unhappy events the Jews finally established a new state in Palestine in the year 1948. They named it Israel, and in 1949, at the first meeting of its 'Parliament', Dr Weizmann was elected the First President of the State of Israel. Such was his well-earned reward.

Lloyd George gave great credit to Weizmann for his part in the negotiations leading up to the Declaration. The chemist's brilliant work in producing acetone, he said, had brought him into 'direct contact with the Foreign Secretary', and 'This was the beginning of an association, the outcome of which was the famous Balfour Declaration, which became the charter of the Zionist movement. So that Dr Weizmann with his discovery not only helped us to win the war, but made a permanent mark upon the map of the world'.[9]

Weizmann's contribution was certainly very great but there were, of course, many other important Jews who were continually pressing the government to restore Palestine to their race. There were also many other reasons for the Declaration than that of rewarding a scientist for his brilliant work.[10] Even before the war a few British statesmen had shown themselves sympathetically disposed towards Zionism and Lloyd George himself respected the Jews as a race. This sympathy increased during the war. Moreover, by 1917 many Allied statesmen had realised the importance of enlisting Jewish support in neutral countries for their war effort, and notably that of the big Jewish population in America. Furthermore, Lloyd George, and other British statesmen, believed that a friendly population of Jews in Palestine would help to guarantee the security of the approaches to the Suez Canal.

There can be no doubt, however, that Weizmann's scientific achievement in manufacturing acetone gave him 'a friend at court'. His association with the Foreign Secretary, Mr Balfour, thus came at a very opportune time for Jewry and Weizmann made good use of the opportunity.

These words of Lloyd George make a fitting end to this story of a brilliant Jewish chemist and statesman: 'The rebuilding of Zion is the only reward he seeks and his name will rank with Nehemiah in the fascinating and inspiring story of the Children of Israel'.

21. *A Jewish Chemist is cast out*

THE PREVIOUS story tells of the reward gained by a Jewish chemist during World War I for his work for Britain and her Allies. A most distinguished chemist helped the German war effort in the same war. But his ultimate 'reward' was expulsion to a foreign land. This story although about war, begins with the peaceful occupation of farming.

A growing plant takes many substances out of the soil and these are replaced, on cultivated land, by the addition of manure and artificially made fertilisers. In the manufacture of many of these fertilisers, huge quantities of the liquid called nitric acid are used each year.

Up to the beginning of this century most of the nitric acid was made from a white substance called saltpetre which is found especially in Chile and other South American countries. But in 1898 the British chemist, Sir William Crookes, pointed out that such large quantities of this salt were being used that there would soon be little of the natural supplies left. He suggested that chemists should try to discover new ways of making the acid.

This acid is a most important substance; in addition to its use in the manufacture of fertilisers for the farmers it is also used in making explosives. Obviously therefore a country which in peacetime has many factories making it can when war breaks out quickly switch them over to making explosives. The possible double use of these factories may well have quickened the search for new methods of making nitric acid.

In most European countries factories using large quantities of substances which had to be imported from South America could soon be made idle in war-time. For their enemies, by blockading their ports and attacking their ships at sea, could greatly cut down their supplies.

As soon as war broke out in 1914 the Allied navies blockaded Germany and her supplies of saltpetre from South America almost ceased. Germany would soon have been disastrously short of explosives but for the capture of a ship at Antwerp and the work of her chemists.

It happened that a ship loaded with many thousand tons of saltpetre had docked at that Belgian port shortly before war was declared. In the first few days of the fighting the Germans over-ran Belgium and on reaching Antwerp found the ship still fully loaded. For some reason or other those in authority had not sent her to sea nor sunk her nor thrown the cargo into the water. Instead, the ship was left in port loaded with a large quantity of a most important war-time commodity. A well-known chemist has estimated that but for this supply the German stocks of salt-petre would have been exhausted by the spring of 1915.[1]

For some years before the war chemists in Germany and other countries had been searching for methods of making fertilisers from the free and unlimited supply of nitrogen in the air. By 1914 three methods had been discovered but only one of them need be mentioned in this story. This is the one which was discovered by Fritz Haber, the son of Jewish parents of German nationality.

Haber had succeeded in making fertilisers using mainly water and air, and needed nothing from abroad. By 1914 he was actu-ally manufacturing fertilisers, and could readily turn the factory over to making nitric acid. But the quantity produced was very small indeed compared with that which was then being obtained from saltpetre.

On the outbreak of war the German leaders realised the great importance to their war effort of Haber's manufacturing process and quickly built many new factories. Thus by the summer of 1915 Germany was manufacturing large quantities of nitric acid and was rapidly becoming independent of supplies of saltpetre.

Haber had rendered a great service to the country of his birth and the leaders of his nation regarded him as one of their best chemists.

Early in the war the fighting had taken a most unexpected turn. Before its start the war-leaders of both sides had expected that the next war would be one of movement with the infantry and cavalry manœuvring over a wide range of country. But after the first few weeks the fighting developed into trench warfare and new methods and weapons were required. The British invented the tank, as explained in a later book, and the Germans introduced poison gas.

The War Ministry in Berlin, in deciding to investigate the possibility of using such a gas, were probably influenced by letters from soldier-chemists in the trenches strongly advocating its use. The Ministry consulted Professor Nernst of the University of Berlin and he agreed to carry out an investigation. Towards the end of 1914 Professor Haber was given a share in the work and soon took complete charge of it.

There were not many poisonous gases which could be used in warfare because a suitable one had to possess many distinct properties. Ideally it had to be poisonous enough either to kill a soldier or else to put him quickly out of action; if it would not do either of these, it had at least to put him temporarily at a great disadvantage so that his enemy, being protected from the gas, could easily deal with him. It had to be heavier than air so that when it was set free it would not immediately rise to a height of more than six feet, where it would have no effect on a man. This heaviness was an important property in trench warfare, because a heavy gas travelling at ground level falls into trenches and dugouts just as a stream of water would do.

The ideal gas would have had no odour or colour, so that its presence would not have been detected until it had done its deadly work. It was preferable that it should not dissolve in the rain or decompose in the hot summer weather. Moreover there had to be a method of making large quantities of it from materials obtainable even in a blockaded country. Finally it had to be easy to transport.

The experimental work done by Haber and his assistants led him to recommend the use of chlorine, but whether he was the first to suggest this gas is not known. There is no doubt, however, that he favoured its use, and he never sought, even after the war, to disclaim responsibility for its introduction.

Chlorine is made from common salt of which Germany had supplies in plenty. It can be stored in cylinders and so is readily transported. This gas is between two and three times as heavy as air and therefore will sink into shell holes, trenches and dug-outs. It is so deadly that a little of it will kill a man or at least put him out of action for a long time. But on the other hand it is green and has a strong smell, so that its presence is easily detected.

Haber first suggested that the gas should be put into shells but as all available shell cases were required for explosives his alternative suggestion that the gas should be liberated from cylinders at a time when the wind was blowing towards the Allied trenches was adopted instead.

In 1915 Haber was still a civilian and had no high standing in the military world of Germany. Although like all fit Germans he had served his period of conscription he had been posted to the Reserve with only the non-commissioned rank of sergeant some twenty years before the war broke out. For before 1914 a Jew had but a small chance of a commission in the Prussian army. Thus Haber, who to the German generals was merely a civilian, and a Jew at that, could not expect the high-caste, aristocratic leaders of the German General Staff to pay much attention to him.

After much hesitation the Supreme Command agreed reluctantly to a trial at the Front and chose the Ypres Salient as the spot for it. The planning and the direction of this first attack seems to have been left largely to Haber, despite his lack of experience in large-scale military operations.

About 170 tons of chlorine, under pressure, were put into some 5,700 cylinders. The cylinders were taken to the Front and 'dug in' along a three-and-a-half-mile stretch, and the men who were to release the gas were suitably protected by masks. Early in 1915 all was ready for it to be released whenever the wind was favourable.

The gas was set free on 22 April 1915[2] – at a place where the British line joined the line held by French coloured troops from Algeria. The man on look-out saw a greenish cloud followed by white smoke, about a yard high, moving towards him. The gas reached the trenches and sank down into them. There were shrieks of terror; at first the men's eyes, noses and throats began to smart; but before a few minutes had passed many of them began to cough harshly and then vomit blood. Panic set in and most of those who were able to do so left the trenches and rushed to the rear.

The first gas attack in 1915

The Germans then attacked successfully and much ground was gained during the day. But they halted and 'dug in' at half past seven to rest for the night. Had they but known it, the way to Ypres was open to them, for there was a gap of five miles in the Allied front line through which thousands of Germans could have marched during the night.

The halt gave the British time to rush up troops to fill the gap and thus to stop any further advance next day. But it had been a disastrous time for the Allies. Five thousand soldiers had been killed and fifteen thousand gassed; six thousand men had been captured together with fifty-seven artillery guns and fifty machine guns.[3]

Anger and horror ran through the whole British nation; but the main criticism which the government of this country had to face was that no preparations whatever had been made to protect the men against this deadly weapon, although more than one warning had been received that the Germans were preparing such a new method of attack.[4] All these warnings had, it seems, gone unheeded.

Fortunately for the Allies, the German High Command, under General von Falkenhayn, lacked the imagination or foresight to see the value of this new means of warfare and looked on it only as an experiment. Sufficient reserves had not been provided in case it proved successful, and no special tactics for the new weapon seem to have been thought out; at any rate no tactical instructions were issued. It has been stated, however, that the German commanders were not enthusiastic about its employment because its success depended on a suitable wind, and, since this is a very uncertain factor in Flanders, troops might have had to be kept in position for a long time, awaiting a favourable moment.

* * * * *

This story has many features of interest. It deals with the first use on a large scale of a new war-weapon. The early twentieth-century soldier was as loud and vigorous in his protest over the use of this new weapon as were the knights of old when the new weapon, gunpowder, was first used against them (chapter 4); or, for that matter, as the people of 1945 were about the first use of the atomic bomb (which is discussed in a later book).

The story also shows that the German military planners had made no preparations to exploit the great opportunities presented to them on that night in April 1915, when the gas was released for the first time. Evidently they had no great faith in the new chemical method. Had they but realised it, they had a weapon of surprise equal to the British tank which was ready for use in battle a short time afterwards. It is most noteworthy that the leaders of the armies of both Germany and the Allies failed to realise the value of surprise in the first use of these two scientific weapons.

And this was not the Germans' only mistake. In introducing cloud-gas warfare they were cutting their own throats. The use of poison gas in clouds could never have been of permanent advantage to them because on most days of the year the wind in Flanders blew from the Allied trenches towards the German lines. The prevailing winds therefore conferred the advantage in the use of cloud gas on the Allies and not the Germans.

Another interesting feature of the story is the almost unbelievable rise in military circles of Professor Haber. He was promoted quickly to a newly created post of Chief of the Chemical Section of the War Ministry with the rank of a Royal Prussian Captain. He took his orders directly from the Minister of War and Field Marshals Hindenburg and Ludendorff[5] (chapter 22). Very few men, even of pure German descent, have risen from the rank of a junior non-commissioned officer in the Reserve to such an exalted military rank as his then became.

After the war many people in Allied countries regarded Haber with execration and accused him of having committed an unforgivable crime against humanity.

There was a melancholy sequel for Professor Haber. During the years after the German defeat in 1918, times were difficult for all Germans. But gradually Haber got the young scientists together again, and, after the lapse of a few years, was in charge of large laboratories where scientists were busy on research. By 1930 he had attained world-wide fame for his many other brilliant discoveries in science. His work for the Fatherland during the war and peace had surely earned him fame, honour and reward from his fellow countrymen; his future seemed assured.

Then, in the early nineteen-thirties, came the Nazis' rise to power in Germany under Hitler. Hitler's aim was to have a united nation. His teaching was that the Germans were a master race. He roused national emotions and passions in many ways, one being to stir up racial hatred against all who were not of the so-called pure German ('Aryan') stock. The Jews, in particular, were made the victims of this racial hatred.

They were persecuted and maltreated in almost every conceivable way, and by the year 1934 many of them had been forced to flee the country. Many of the wealthy ones who remained were imprisoned and all their property was destroyed or confiscated.

At first, in the early nineteen-thirties, Haber protested against this treatment, but before long this great German chemist, German soldier and German patriot became simply 'the Jew Haber'– nothing more. Nazi Germany had no gratitude for his former work for his country. He became an outcast; no longer fit to live in Germany. Like many other Jews he sought refuge abroad. A sick man, he first went to a sanatorium in Switzerland. Then England offered him a home, and he was invited to live in Cambridge, where he enjoyed the hospitality of the University Chemistry Laboratories for a short time. But the strain of the previous years had been too great for him and he died of a heart attack in Basle in January 1934.[6]

22. Alchemy at intervals of three hundred years

ALCHEMY HAS been defined as the chemistry of the Middle Ages. Most of the alchemists concentrated their efforts on converting base metals such as mercury and lead into gold and silver. There was the centuries'-old belief that the gold and silver found in the ground, had 'grown' gradually in the course of thousands of years from the baser metals. The alchemists tried to accelerate this natural process in the laboratory.

Many of them were serious students of their subject and made valuable contributions to man's knowledge of chemistry. Some of them tried to find the elixir of life, a universal medicine which would cure all ills and ensure long life to those partaking of it. But there were some among them who were downright rogues.

In practice, the method followed by almost all the alchemists was to try to find something which, when added to a base metal, would produce gold. This something was given the name of the Philosopher's Stone, and the process by which gold would ultimately be produced was called transmutation.

Three stories, dated at intervals of about three hundred years, have been chosen from numerous others to illustrate that the belief in the transmutation of base metals into gold has been a most persistent one and is even believed by some people today. The first story describes an incident which occurred in the year 1329; the next, an experiment performed about three hundred years later, and finally comes an account, taken largely from a national newspaper, of the trial of an alchemist who was arrested in 1929 – a date exactly six hundred years after the first incident mentioned in this chapter.

* * * * *

In the year 1329 England was ruled by Edward III who, like most monarchs of his time, was always short of money. On hearing of what seemed a marvellously easy way of getting some he issued the following order:

> Know all men that we have been assured that John Rows and William de Dalby know how to make silver by the art of alchemy, that they have made it in former times and still continue to make it; and considering that these men, by making that precious metal, may be profitable to us and to our kingdom, we have commanded our well-beloved William Carey to apprehend the foresaid John and William wherever they can be found and bring them to us together with all instruments of their art under safe and sure custody.[1]

Unfortunately for the King's exchequer, his well-beloved William could find neither his namesake nor John; and nothing further seems to have been heard of them.

The same king of England, according to John Cremer, an 'Abbot of Westminster and Brother of the Benedictine Order', suffered another disappointment from another alchemist, the famous Raymond Lull.

Lull, a Spanish nobleman born about 1235 A.D., became a Franciscan monk of great renown. He claimed that he possessed

The Alchemist in his laboratory

the philosopher's stone, 'a morsel of precious medicine as big as a bean', and that by its use he could turn mercury into a gold that was purer than the gold obtained from mines.

Such was the reputation of this alchemist that many believed him. It is said that he visited many countries, and this description of his visit to England is given by Cremer in *The Testament*.

I introduced my noble master to his most gracious majesty King Edward who received him kindly and honourably and obtained from him a promise of inexhaustible wealth on condition that he (the King) should in person conduct a crusade against the Turks, the enemies of God, and that he should thenceforth refrain from making war on other Christian nations. But alas, this promise was never fulfilled, because the King grossly violated his part of the contract and compelled my dear master to fly beyond the seas with sorrow and grief in his soul.[2]

There are some other versions of the reputed visit of Lull to this country. Some state that he came in Edward III's reign, others put the date as 1312, when Edward II was on the throne, whilst others, again, doubt if he ever came at all.

It is said that during his stay he was lodged either in a monk's cell in Westminster Abbey or in the Tower of London. According to many accounts, he was successful in making gold out of iron, quicksilver and lead to the value of six million pounds. Some of this gold was used, reputedly, to make golden coins which were given the popular name of Raymond's nobles. (Some of them, according to tradition, were in existence a few centuries later.) It is also said that gold dust was found on the floor of Raymond's cell many years after he had fled the country, in disgust, when the King broke his promise and went to war against Christian France.

* * * * *

The attempt to arrest John and William is authentic since the order for their apprehension appears in the Patent Rolls for the year 1329. But the story told by Abbot Cremer about Raymond Lull is now discredited. It would appear that no one of the name of John Cremer was Abbot of Westminster in the reign of either of the two Edwards, and therefore Cremer's *Testament*, in which the story is related, is thought to be fictitious. Nevertheless, many writers believe that Lull did visit England and helped to coin the nobles. But Edward obtained the gold, it has been suggested, by following Lull's advice to put a tax on wool, a tax which brought thousands of pounds to the Royal Coffers,[3] and not by using the Philosopher's Stone.

* * * * *

A little more than three hundred years later, on 27 December 1666, a stranger named Elias the Artist called upon John Frederick Helvetius, court physician at the Hague to the Prince of Orange, who later became William III of England.

This visit and its consequences were described in a book

written shortly after its occurrence which has this most illuminating title: *The Golden Calf: That most rare Miracle of Nature in the transmutation of metals: viz: How at the Hague a mass of lead was in a moment of time changed into Gold by the Infusion of a small particle of our Stone.*

The writer relates that after a preliminary conversation Elias pulled out of his pocket an ivory box containing three pieces of a glass-like substance, 'the colour of sulphur and the size of a walnut'. These, he said, were pieces of the Philosopher's Stone.

The Philosopher's Stone

Elias allowed Helvetius to hold one of the pieces in his hand and to examine it. Next he showed Helvetius five large gold plates which he claimed had been made from gold obtained by using his Philosopher's Stone. Then he left, promising to return later.

A few weeks later Elias returned and handed to Helvetius 'a small particle as big as a rape seed', saying, 'Take of the greatest treasure of the world which very few great kings or princes could ever see'. To this Helvetius replied that the piece was much too small to be of any real use. Thereupon Elias took it and broke it in two with his thumb nail, threw one of the pieces thus formed into the fire and wrapped the other in blue paper. This he gave to Helvetius, saying, 'It is sufficient for thee'. Helvetius thanked him, and said that he would like to experiment with it next day, if Elias would give him the necessary directions.

He was told to wrap the stone in yellow wax and then to put it into molten lead; the wax would protect the stone from the fumes of lead until it had penetrated far enough into the molten metal. Elias then left promising to return on the morrow. Next day came, but no Elias. Therefore Helvetius decided to do the experiment himself with the help of his wife.

He heated half an ounce of lead in a crucible whilst his wife wrapped the precious stone in wax. When the lead had melted he threw the wrapped stone into it. Immediately there was a hissing noise and an effervescence, and then, after a quarter of an hour had passed, they saw to their great delight that the whole mass of lead had turned into gold. What happened next can be related in the words of Helvetius himself.

> I, and all who were present with me, being amazed, made what haste we could to a goldsmith, who after a precious examination, judged it to be gold most excellent and that in all the world better could not be found, withall adding that he would give me fifty florins for every ounce of the gold.

Helvetius was a man held in high esteem by the Prince and his fellow men and there is no doubt that he himself believed that he had actually transmuted the lead. But there is no record of any other gold being made either by him or by Elias the Artist and the story is now discredited.

In the three hundred years which followed many great advances were made in our knowledge of chemistry, yet even in the twentieth century there were people who believed in the transmutation of base metals into gold, as the next story illustrates.

In the year 1925 a German named Franz Tausend announced that he had been successful in transmuting a base metal into gold. This, he said, had been achieved because of an accidental explosion which occurred during one of his experiments which had nothing to do with transmutation. Tausend stated that his theory was based on the belief of the ancients that gold grew in the ground by a natural but extremely slow process of growth from base metals. He claimed that he could do in a few hours what had taken Nature hundreds of thousands of years to achieve.

Tausend approached the authorities of the Munich Mint with his astounding news, but they rebuffed him. He then got in touch with Ludendorff who had been one of Germany's greatest generals in World War I (page 126), and had later been elected a member of the German Parliament (the *Reichstag*).

Ludendorff decided to investigate Tausend's claim and appointed his stepson to help in the inquiry. Two years later the stepson declared that while investigating the claim he had become convinced that Tausend had discovered the secret of transmutation. He said that he had seen forty to fifty experiments done under close supervision and that in most of them Tausend was successful in obtaining a piece of gold the size of a pin's head.

The stepson also affirmed that Tausend had shown him part of the process of making gold but had not disclosed the final stage. Nevertheless, he added, some of the experiments made in Tausend's absence had seemed to yield very satisfactory results. He had satisfied himself in 1927 that there was no possibility of fraud.

At first General Ludendorff was extremely sceptical and continually demanded repetitions of the experiments. But after receiving the report of the investigations he became convinced that Tausend had discovered the age-long secret and instructed his legal representatives to form a company to exploit the method.

Most of the shareholders of the company were people who were closely associated with the general, and they included members of many of the leading and noble families in Germany. It has since been said that these well-known people did not take up shares solely in order to make a great deal of money for themselves, but also to break the sway of capitalism. For Germany was then passing through a most difficult period financially and these shareholders believed that her economic position would be greatly improved by a decrease in the value of gold. Hence 75 per cent of the company's expected profits were to be given to General Ludendorff 'for patriotic purposes' and 20 per cent to be divided amongst the shareholders, whilst Tausend's reward was to be the remaining 5 per cent.

By the year 1928 Tausend was living under the name of Baron Tausend in a princely palace, enjoying life on the capital of the company. But this extravagant way of living did not last long, for before the end of 1929 he was arrested on a charge of fraud.

The trial commenced at Munich in January 1931. Evidence was first called about events up to and including the formation of the company. In addition to what has already been mentioned in this chapter the evidence included the statement of a Hamburg manufacturer who produced in court a piece of gold the size of a pin's head which, he said, Tausend claimed to have made. A director of the Deutsche Bank told the court that he had seen a lump of gold the size of a duck's egg which Tausend also claimed to have made; and Tausend's business manager said that he had seen Tausend apparently produce twenty grams of gold during an experiment, declaring that 'a cold shiver went down his spine when he saw the gold'.

Evidence was also given that Tausend had shown other people pieces of gold which he claimed to have made by the process of transmutation. A former employee of the company gave evidence that on one occasion he had discovered in a cupboard in the laboratory a test tube filled with gold dust, and that Tausend had explained that the gold was used in the process because the addition of a small quantity of gold to the new material greatly facilitated the transmutation.

Expert chemical witnesses were then called to give evidence; and all of them stated that Tausend had failed to produce gold in experiments done in their presence.

One of these expert witnesses, a director of the Mint, told the Court that he had watched Tausend perform an experiment with apparent success, and was very much surprised when he saw gold in the melting pot at the end of it. He added that he later became suspicious when he remembered that Tausend carried a fountain pen with a gold nib. He therefore analysed the 'gold' produced by Tausend in this experiment and found that it was an alloy like that used by manufacturers of nibs for fountain pens!

Tausend was found guilty and sentenced to a term of imprisonment of three years eight months. He was also ordered to pay the cost of the trial, and all the materials found in his laboratory, including some gold, were confiscated. The sentence, said the President of the court, was a light one, considering that a very large amount of money had been obtained by a great cheat; but the court could see its way to giving a fairly light punishment because those who had lost their money had been only too ready to believe that they were going to get a fortune in an easy way.[4]

* * * * *

There is this to be said about the ease with which Tausend deceived many well-known people. Recent advances in the study of the atom had shown that transmutation, of a kind, was not impossible. Long before the year 1925, when he first began to exercise his persuasive powers, scientists had abandoned the age-old theory that an atom cannot be divided. The discovery of radioactivity in 1896 (see Book II), had led to the knowledge that the atoms of a few elements spontaneously undergo a change by natural process; and by the year 1920 Rutherford had succeeded in splitting a few atoms by bombarding them with particles which he had obtained from a natural source.

Therefore a rogue who could tell a good story was more likely to be believed in 1925 than he would have been just before 1896.

It is of interest to note that while Tausend was serving his prison sentence two Cambridge scientists, Cockcroft and Walton, succeeded in splitting atoms by an artificial method in the laboratory, and that since 1932 numerous physicists have carefully studied the nature and properties of atoms with such success that, today, the atoms of many elements can be split and transmutation of some elements is thereby almost a daily occurrence.

But no one has yet achieved the aim of the alchemists – the transmutation of base metals into gold.

References, Chapters 1-4

CHAPTER 1 (pp. 9–16)

1. Pliny, XXXVI, 64.
2. G. W. Morey, *The Properties of Glass* (1954), p. 4.
3. B. Palissy, *Traité des eaux et fontaines* (1580), p. 156.
4. Sir Flinders Petrie, 'Glass in the Early Ages', *Journal of the Society of Glass Technology*, vol. X (1926), pp. 229–34. Also J. R. Partington, *Origins and Developments of Applied Chemistry* (1935), p. 120.
5. Dion Cassius, *Roman History*, 57, 21, 7.
6. Pliny, V, 33.
7. Titus Petronius Arbiter, *Satyricon*, cap. 51.
8. J. R. Vavra, *5000 Years of Glassmaking* (1955), p. 40.
9. J. H. de Blancourt, *Art de la verrerie* (1699), pp. 13-14.
10. J. Wilson, 'Safety Glass', *Journal of the Society of Glass Technology*, vol. XVI (1932), p. 67.

CHAPTER 2 (pp. 16–22)

1. Polybius, III. Also Livy, XXI.
2. Polybius, III, 54. Also Livy, XXI, 32.
3. Livy, XXI, 37.
4. J. Watson, *Chemical Essays* (1794), vol. I, p. 546.
5. Marcellinus, XV, 10. Also Appian, *Wars of Hannibal*, I, 505.
6. *Notes and Queries*, 4th Series, vol. II, pp. 289 and 350.

CHAPTER 3 (pp. 22–27)

1. Plutarch's *Lives* (Antony).
2. Pliny, IX, 35.
3. Pausanias, VIII, 18.
4. Pliny, IX, 35.
5. L. Dutens, *Origins of Discoveries* (1794), p. 253 ff.
6. G. Pancirollus, *A History of Many Memorable Things* (1715), vol. I, p. 20.
7. J. W. Burgon, *The Life and Times of Sir Thomas Gresham* (1839), vol. II, p. 354.

CHAPTER 4 (pp. 27–31)

1. G. Pancirollus, *op. cit.* vol. II, p. 384.
2. S. Butler, *Hudibras*, Grey's notes (1774), vol. I, p. 128. Also Polidore Vergil, trans. J. Langley (1663), p. 91.
3. De Blainville, *Travels Through Holland, Germany, etc.* (1743), vol. I, p. 218.
4. H. Wilkinson, *Engines of War* (1847), pp. 137–8.
5. O. Guttmann, *The Manufacture of Explosives* (1895), vol. I, p. 16.

6. H. Wilkinson, *op. cit.* p. 47.
7. Polidore Vergil, *op. cit.* p. 91.

CHAPTER 5 (pp. 32–35)

1. Thomas Thomson, *A History of Chemistry* (1830), vol. I, pp. 75.
2. *Ibid.* vol. I, p. 44. Also E. von Meyer, *A History of Chemistry*, trans. G. McGowan (1888), p. 53 ff.
3. Samuel Johnson, *A Dictionary of the English Language* (1786).
4. H. Kopp, *Geschichte der Chemie*, trans. M. McGowan (1847), vol. IV, p. 144.
5. B. Valentine, *The Triumphal Chariot of Antimony*, trans. T. Kerkringus (1893), p. 74.
6. E. von Meyer, *op. cit.* pp. 34-5.

CHAPTER 6 (pp. 36–42)

1. J. Beckmann, *History of Inventions and Discoveries* (1846), vol. I, p. 191.
2. L. Charlton, *The History of Whitby* (1779), p. 305.
3. W. Camden, *Britannia* (1806), vol. II, p. 11.
4. L. Charlton, *op. cit.*
5. Abraham Rees, *Cyclopaedia* (1819).
6. J. Aubrey, *Lives of Eminent Men* (1813), vol. II, p. 281 ff.
7. G. Young, *A History of Whitby* (1817), p. 807.

CHAPTER 7 (pp. 42–49)

1. R. Hakluyt, *The True History of the Conquest of New Spain*, trans. A. P. Maudsley (1906), vol. I, p. 209.
2. F. H. A. von Humboldt, *Political Essays on the Kingdom of New Spain*, trans J. Black (1811–12), vol. III, ch. XII.
3. Hakluyt, *op. cit.*, vol. I, p. 287.
4. W. H. Prescott, *History of the Conquest of Mexico* (1886), p. 253.
5. *Ibid.* p. 253. Also Humboldt, *op. cit.* vol. III, ch. XII.
6. Prescott, *op. cit.*, p. 253.

CHAPTER 8 (pp. 49–51)

1. H. Pownall, *Some Particulars relating to the History of Epsom* (1825).
2. N. Grew, *A Treatise on the Nature and Use of the Bitter Purging Salt* (1697), p. 9.

CHAPTER 9 (pp. 52–55)

1. *The Gentleman's Magazine*, vol. XV, p. 310.
2. R. Hurd, *Joseph Addison's Works* (1854), vol. I, p. 436.

3. De Blainville, *Travels Through Holland, Germany, etc.*, trans. G. Turnbull and W. Guthrie (1745), vol. III, pp. 383 ff.

CHAPTER 10 (pp. 55–59)

1. T. Thomson, *History of Chemistry* (1830), vol. I, p. 328. Also E. von Meyer, *History of Chemistry*, trans. G. McGowan (1891), p. 151.
2. E. Grimaux, *Lavoisier (1743-94)*, (1888), pp. 322-3.
3. T. E. Thorpe, *Essays on Historical Chemistry* (1894), p. 107.
4. T. Carlyle, *The French Revolution* (Century ed.), vol III, p. 263.
5. *Calendar of the Lycées des Arts*, Year IV, p. 89.

CHAPTER 11 (pp. 59–66)

1. *Philosophial Transactions* (abridged), vol. VII, pp. 4–6. Also J. Bancroft, *The Philosophy of Permanent Colours* (1813), vol. II, p. 60.
2. A. J. Greenaway, *The Life and Work of Professor W. H. Perkin*, The Chemical Society, London (1932), p. 8.
3. R. Robinson 'The Life and Work of Sir William Henry Perkin', *American Dyestuff Reporter* (1956), vol. XLV, p. P.760.
4. W. H. Perkin, Hofmann Memorial Lecture, *Journal of the Chemistry Society* (1896), vol. LXIX, p. 604.
5. *Ibid.* p. 604.
6. R. Robinson, *op. cit.* p. P.760. Also R. Meldola, 'W. H. Perkin', *Journal of the Chemistry Society* (1908), vol. XCIII, p. 2214.
7. *Encyclopaedia Britannica*, 9th Ed., vol. II, p. 48.

CHAPTER 12 (pp. 66–74)

1. Hatton Turnor, *Astra Castra* (1865), p. 41.
2. *The Annual Register*, vol. XXVI, p. 65 ff.
3. Dispatch from Paris, 5 Sept. 1783, quoted from *Haude-Spenersche Zeitung*, Berlin, no. 113.
4. Hatton Turnor, *op. cit.* p. 41.
5. B. Franklin, *Writings* (*n.d.*), vol. IX, pp. 155–6.
6. F. M. Grimm, *Correspondance littéraire, etc.* (1877–82), vol. XIII, p. 319.
7. G. Bugge, *Das Buch der Grossen Chemiker* (1929), vol. I, p. 386 ff.
8. Sir William Ramsay, *The Life and Letters of Joseph Black* (1918), pp. 78–83.
9. Thomas Thomson, *A History of Chemistry* (1830), vol. I, pp. 328–9.

CHAPTER 13 (pp. 74–80)

1. *Philosophical Transactions* (abridged), vol. VIII, p. 295–6.
2. S. Smiles, *Lives of the Engineers* – 'Boulton and Watt' (1878), p. 346.

3. Smiles, *Men of Industry and Invention* (1892), p. 136.
4. W. Murdoch, *A Letter to a Member of Parliament* (1809).
5. M. Heaviside, *The True History of the Invention of the Lucifer Match* (1909).
6. Smiles, *Lives of the Engineers, etc.*, p. 346 footnote.
7. Caricature by G. M. Woodward.
8. W. Murdoch, *A Letter to a Member of Parliament.*
9. Smiles, *Lives of the Engineers*, p. 349.
10. E. F. Armstrong, Murdoch Centenary Lecture, *Transactions of the Institute of Gas Engineers*, vol. LXXXVIII, pp. 933–1000.

CHAPTER 14 (pp. 81–87)

1. Quotations from J. Priestley, *Experiments and Observations on the Different Kinds of Air* (1790), part I, section I, and also from J. Priestley, *Memoirs* (1806), p. 61.

CHAPTER 15 (pp. 87–90)

1. W. Haythornthwaite, *Harrogate Story* (1954), p. 23.
2. T. Garnet, *A Treatise on the Mineral Waters of Harrogate* (1894), p. 147.
3. J. Scoffern, *Chemistry No Mystery* (1839), p. 186.

CHAPTER 16 (pp. 90–94)

1. J. Dalton, *Extraordinary Facts Relating to the Vision of Colours* (1794).
2. H. E. Roscoe, *John Dalton and the Rise of Modern Chemistry* (1895), p. 75.
3. W. C. Henry, *Memoirs of the Life and Scientific Researches of John Dalton* (1854), p. 24.
4. H. E. Roscoe, *op. cit.* p. 75.
5. H. Lonsdale, *The Worthies of Cumberland* (1874), p. 99 ff. Also H. E. Roscoe, *op. cit.* p. 75.
6. *Ibid.* p. 73.
7. C. Babbage, *Passages From the Life of a Philosopher* (1864), p. 272.
8. T. Wemyss Reid, *Memoirs and Correspondence of Playfair* (1899), p. 58.
9. E. M. Brockbank, *John Dalton* (1944), p. 45.

CHAPTER 17 (pp. 94–100)

1. F. R. Japp, *The Kekulé Memorial Lecture*, The Chemical Society, London (1897), p. 97.
2. *Ibid.* p. 100.
3. *Ibid.* p. 100.
4. *Ibid.* p. 130.

References, Chapters 18-21

CHAPTER 18 (pp. 100–107)

1. O. L. Erdmann, *Journal für praktische Chemie* (1851), vol LII, 428
2. C. J. Fritzsche, *Mem. acad. St Petersbourg* (1870), 1.7.15.
3. C. L. Mantell, *Tin* (1949), p. 10.
4. L. Huxley, *Scott's Last Expedition* (1913), vol I, p. 173.
5. *Ibid.* vol. I, p. 578 and note on p. 630.
6. *Ibid.* vol. I, p. 592.
7. *Ibid.* vol. I, p. 606.
8. C. L. Mantell, *op. cit.* p. 10.
9. L. Huxley, *op. cit.* p. 631.
10. *Ibid.* vol. II, p. 345.
11. *Ibid.* vol. I, p. 631.
12. E. S. Hedges, *Tin and its Alloys* (1960), p. 60.

CHAPTER 19 (pp. 107–111)

1. *Encyclopaedia Britannica*, 9th ed., vol VII, p. 583 and vol XVII, p. 521.
2. A. Marshall, *Explosives* (1932), vol. I, p. 357.
3. H. Schuck and R. Sohlman, *The Life of Alfred Nobel* (1924), p. 101. H. E. Pauli, *Alfred Nobel* (1947), pp. 86–90.
4. A. R. Marble, *The Nobel Prizewinners* (1932), p. 5.
5. H. Schuck and R. Sohlman, *op. cit.* p. 132. H. E. Pauli, *op. cit.* p. 138.
6. Vance Thompson, 'The Nobel Prizes', *The Cosmopolitan Magazine*, vol. XLI, p. 468.

CHAPTER 20 (pp. 112–119)

1. C. Weizmann, *Trial and Error* (1949), p. 220.
2. *Ibid.* p. 220.
3. J. L. Hammond, *C. P. Scott* (1934), pp. 196–7.
4. David Lloyd George, *War Memoirs* (1933), vol. II, p. 585.
5. *Ibid.* vol. II, p. 586.
6. E. C. Dugdale, *Arthur James Balfour* (1936), vol. II, p. 233.
7. *Ibid.* vol. I, p. 433, and vol. II, pp. 223–6.
8. Lloyd George, *op. cit.* p. 586.
9. Lloyd George, *op. cit.* p. 587.
10. I. Cohen, *A Short History of Zionism* (1951), p. 74.

CHAPTER 21 (pp. 120–127)

1. J. E. Coates, Haber Memorial Lecture, *Journal of the Chemistry Society* (1935).
2. J. E. Edmonds and G. C. Wynne, *History of the Great War: France and Belgium 1915* (1927), pp. 163–194.
3. C. Wachtel, *Chemical Warfare* (1941), XXVIII, ch. II.
4. *Nature*, vol. CXXXV (1935), pp. 176 and 216.

5. C. Wachtel, *op. cit.* ch. II.
6. Haber Memorial Lecture.

CHAPTER 22 (pp. 127–135)

1. *The Percy Anecdotes.*
2. A. E. Waite, *The Testament of Cremer in the Hermetic Museum*, vol. II, pp. 71–2.
3. C. Mackay, *Memoirs of Extraordinary and Popular Delusions* (1852), vol. I, p. 109.
4. *The Times*, various issues between 22 January and 6 February 1931.